MW00576371

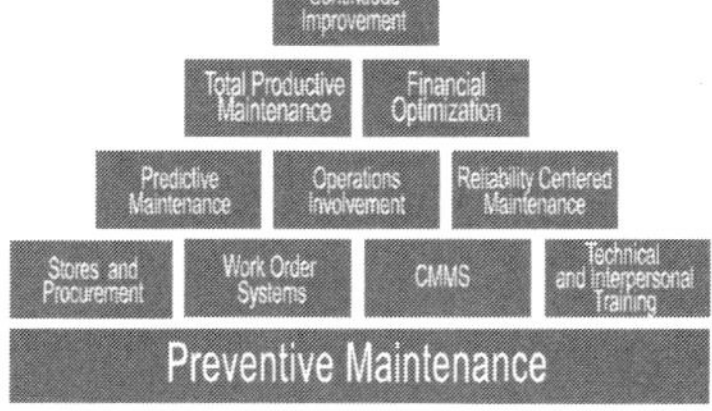

MRO Inventory and Purchasing

Terry Wireman, CPMM

www.terrywireman.com
TLWireman@Mindspring.com

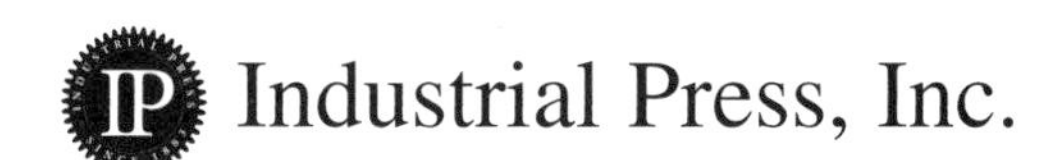

Library of Congress Cataloging-in-Publication Data

Wireman, Terry.
MRO inventory purchasing / Terry Wireman.
p. cm. -- (Maintenance strategy series)
Includes index.
ISBN 978-0-8311-3329-0 (hardcover)
1. Plant maintenance. 2. Inventory control. I. Title.
TS192.W565 2007
658.2'02--dc22

2007043212

Industrial Press, Inc.
989 Avenue of the Americas
New York, NY 10018

First Edition, 2008

Sponsoring Editor: John Carleo
Interior Text and Cover Design: Janet Romano
Developmental Editor: Robert Weinstein

10 9 8 7 6 5 4 3 2 1

TABLE OF CONTENTS

INTRODUCTION

Volume 2: MRO Inventory and Purchasing

MRO stands for maintenance, repair, and overhaul. In the context of this publication, MRO inventory will include disposable and repairable equipment spare parts and consumable maintenance supplies. A more definitive list of these item types is as follows:

1. Disposable spare parts whose inventory levels are replenished by reordering from a supplier.
2. Disposable spare parts, whose inventory levels are replenished in a fabrication shop (internal or external).
3. Spare parts that can be rebuilt and returned to inventory after the repair.
4. Critical spares that need to be monitored periodically to guarantee they are in prime condition when needed.
5. Shop items such as steel, sheet metal, or tubing.
6. Maintenance supply items.
7. Manufacturing and general plant supply items such as janitorial supplies, clothing items, etc.
8. Tools and maintenance test equipment.
9. Parts and supplies waiting to be used or project activities.
10. Excessive parts and equipment left over from projects and awaiting resolution (stock, return, or disposal)

Although this information seems very straightforward, it is surprising to find the current status of MRO spare parts management. In most organizations, the storeroom provides poor service levels. This results in hoarding and pirating of spare parts. Such hoarding, of course, will increase the dollar value of the spare parts a company has on hand. In addition, there are many inaccuracies in the stock counts of items on hand in the storeroom. These inaccuracies result in either having too many spare parts on hand, or delays in work execution due to having too few spare parts on hand. One final problem that is encountered in larger companies is the lack of tracking of the items in remote storerooms. This problem typically leads a company to carry excessive inventory, which increases the overall MRO costs for a company.

Each of these problems has an impact on the efficiency of the maintenance organization. It is not that these conditions are deliberate. They occur because personnel are trying to circumvent the MRO controls. The maintenance personnel try to circumvent these controls because they see them as an obstacle to keeping the equipment operating properly.

From an inventory control perspective, it is necessary to have the right parts, the right number of parts, and properly-managed inventory. This control is necessary if the inventory and purchasing organizations are going to support the equipment maintenance department. Proper controls are important because spare parts cost may make up 40–60% of the total maintenance budget. For this reason, inventory management for MRO spare parts must be given close consideration.

Figure I-1 shows the relationship between maintenance materials and plant profitability. As can be seen, materials typically have the focus of being able to decrease expenses. Of course, decreased expenses will increase profits because expense dollars not spent drop to the bottom line. Therefore, controlling MRO material costs is a quick way to increase profits as some companies see it. This view is shortsighted, however, because decreasing MRO material costs finally will reach a level that decreases profitability, due to increased equipment downtime and reduced capacity.

The impact that MRO parts can have on capacity is highlighted by looking at the left side of Figure I-1. The left side of the diagram shows that increased capacity is achieved by increasing availability or increasing the efficiency of an asset. If the proper MRO spare parts are not available when needed, increased down time of the equipment can result. This added down time, of course, will reduce availability. If the equipment cannot meet the production schedule, then either late shipments or increased cost such as overtime to produce the product will be incurred. Both of these situations will decrease profitability.

Therefore, it can be seen that a proper balance needs to be achieved. Having too many spare parts makes it difficult to control expenses, whereas having too few spare parts makes it difficult to achieve the proper amount of equipment capacity. Consider Figure I-2, which lists typical

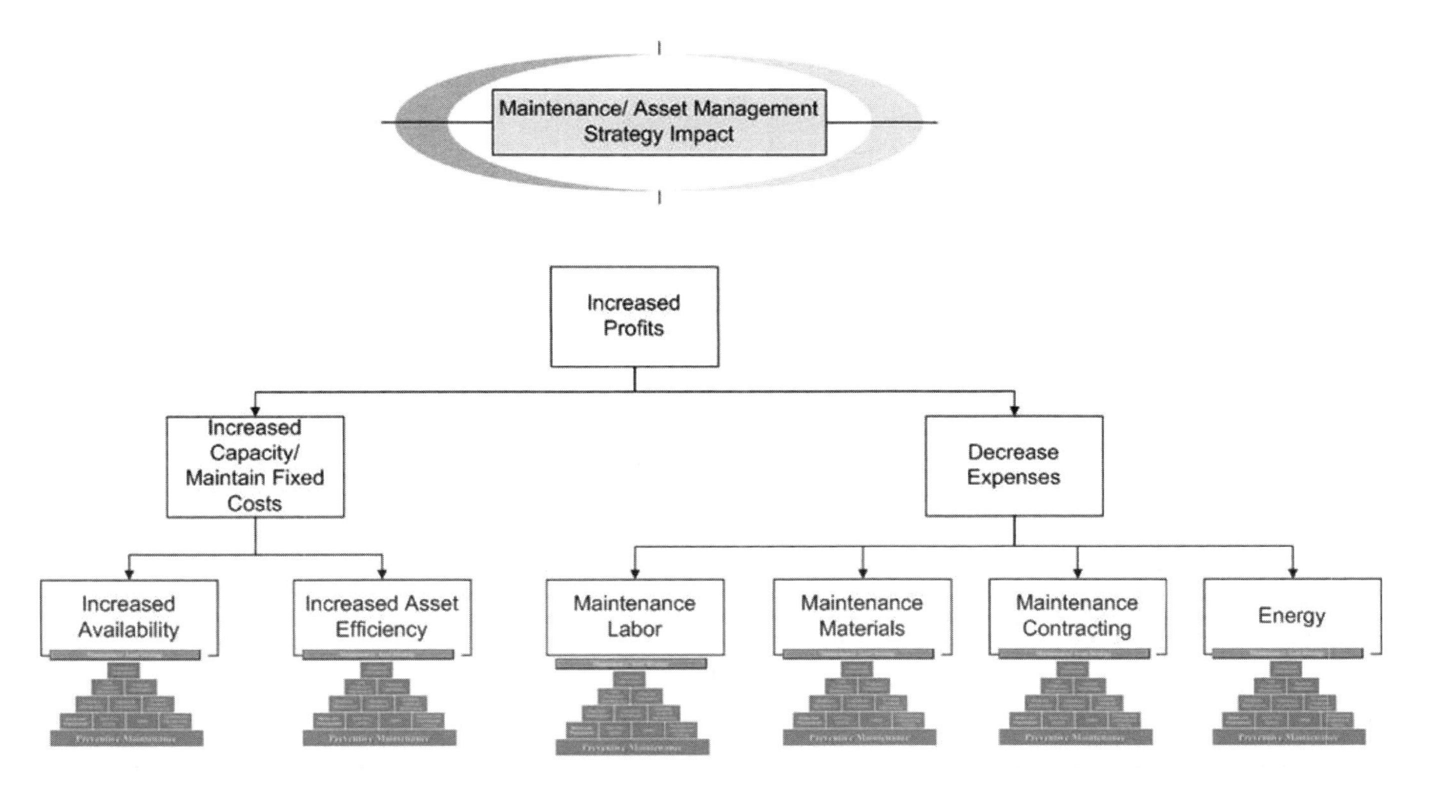

Figure I-1 Maintenance / Asset Management Strategy Impact

material-related delays for maintenance technicians. Consider the amount of time maintenance technicians spend waiting on materials, traveling to get materials, and transporting materials to the job site. These delays can consume a considerable amount of time. When dealing with maintenance technicians, time is money.

Material-Related Waste

- Waiting on materials
- Travel time to obtain materials
- Time to transport materials to job site
- Time to identify untagged materials
- Time to find substitute materials
- Time to find parts in remote/ alternative locations
- Time to obtain Purchase Order approvals
- Cost of processing Purchase Order
- Time lost due to:
  - Other crafts having material problems
  - Wrong materials planned, ordered, or delivered
  - Materials out of stock

Figure I-2 Material-Related Waste

The role of time and, therefore, productivity, is highlighted in Figure I -3. In looking at the wrench time for a technician, it can be seen that if a technician works at a 30% rate, versus a 60% rate, the dollar value can be considerable. What makes the difference and a 30% wrench time, versus a 60% wrench time? It is the amount of delay a technician experiences in attempting to perform a particular job. If, as depicted in Figure I–2, technicians encounter multiple or lengthy material related delays, their wrench time will be significantly lower. Depending on the actual hourly rate of a maintenance technician, significant material-related delays can greatly increase the overall maintenance expenses.

What about other material-related wastes and MRO inventory and purchasing? Again referring to Figure I-2, it can be seen that there may be unidentified materials that will not be utilized in the storeroom. Or there may be parts located in remote or satellite storeroom locations that are not

Productivity Impact
A Typical Example

Reactive	Best Practice
100 Technicians X 2000 Hours/year	100 Technicians X 2000 Hours/year
200,000 hours paid for x 30% Productivity	200,000 hours paid for x 60% productivity
60,000 hours	120,000 hours
Difference of 60,000 additional hours	Difference in dollars 60,000 x $20.00= Potential of $1.2 million

Figure I-3 Productivity Impact

properly tracked. Either of these situations will increase the total valuation of the inventory on hand. If a company has invested expense dollars in spare parts, then any spare parts that are not required or are excess are wasted dollars that could be converted to profit dollars. Again, this potential for waste provides a very good reason for more tightly controlling MRO inventory.

Another point that can be taken from Figure I-2 is the delay time it takes to process a purchase order. Suppose the MRO stocking levels are set incorrectly and it is a lengthy procedure to process a purchase order. If a stock out occurs, then a large period of equipment down time will be incurred. Two thoughts can be derived from this. First, it is necessary to have the proper maximum / minimum stock levels set for all spare parts in order to avoid any stock outs. Second, the purchase order approval process needs to be as streamlined as possible. This will reduce the delay time in procuring a critical item if a stock out does occur.

A final point that can be derived from Figure I -2 is the cost of processing a purchase order. This cost is an internal one for most companies—to process the paperwork necessary and gain the approvals necessary to produce a purchase order. This cost can range in some companies from $35 per purchase order, which is fairly efficient, to well over $200 per purchase order. In large municipalities that require multiple sign offs, some purchase order costs may exceed $300 per purchase order.

There are two areas to address that can lower the cost of processing a purchase order to buy a part. The first is to consolidate items on a single purchase order. Although this is not a direct one-to-one reduction, there is a reduction in total processing costs, when multiple line items are on a single purchase order. The second area where the cost can be lowered is by reducing the number of approval levels to process the purchase order. Each individual who is involved in the process adds costs to the process; therefore, having fewer individuals involved will lower the overall cost.

One last point on the costs to procure materials is related to reactive maintenance organizations, which have a tendency to expedite many spare parts. The expediting costs can become extremely high. Instead of being able to use traditional ground shipment, expensive overnight air shipments are utilized, raising the expense to procure materials need for short term, unplanned work. In some cases, the overnight shipments become a "crutch" for an organization. In one case, there were overnight shipments in one stores location that had been there for over a week and were still in their boxes. This example shows that reactive organizations do not look far enough in the future to avoid these costs—nor do they even understand the waste they are creating. The fact is that the more proactive an organization can be, through the use of their preventive maintenance program, the lower the expediting costs will be for spare parts.

What financial components actually make up holding costs and why do organizations use these to justify continual reduction of inventory? The list includes:

- Holding Costs (the cost of money)

This is the cost of the money tied up in the spare parts, when the company could have the money invested and earning a return.

- Storage Costs

This is the cost of storage space, the cost of labor to move, and other costs such as heat, light, and depreciation.

- Insurance and Taxes

This is the taxes on the inventory investment and insurance for the space and the goods

- Losses due to obsolescence, deterioration, and vanished parts

In total, these costs can add up to as much as 30% of the value of the inventory. Such costs provide the underlying reason for companies focusing on reducing MRO inventory. The lower that the total inventory value is, the lower the total holding costs will be. Unless there is a balance between the holding costs and the cost of a stock out (in related equipment downtime and lost production), companies will always "downsize" their inventories. But doing so will ultimately have a negative impact on their ability to make a profit.

In some organizations, the maintenance department is responsible for their own storeroom. This means the maintenance department will stock their own items, order their own items, and develop their own issue and return systems. In other organizations, an inventory group, the purchasing department, or even accounting may be responsible for the storeroom. Does this make a difference in the performance of the storeroom? Typically, it does not. It really does not matter who operates the stores function, whether it is the maintenance department or some other department. What matters is the focus on customer satisfaction and good inventory management practices. This focus ensures that the proper levels of inventory are stocked, and that the proper levels of equipment availability are maintained.

Inventory Tracking Systems

Most companies today have some form of a computerized inventory management system. In the past, some companies used a form of paper system to track their inventory. However, the price of inventory management programs has dropped to the point that even the smallest organization can afford a suitable program. What does it take to comprise an effective inventory management program? There are several modules that will comprise an effective inventory management system.

The first module is the inventory control module. The inventory control module describes each spare part, assigns a specific number to each part, and identifies a piece of equipment where the part is used. In addition, the inventory control module maintains a record for all stock materials, including the vendor, the initial order, and the reorder process. The inventory control module will also track part issues and returns and helps to expedite the physical inventory process.

The second module is a purchasing module. This module is utilized to purchase spare parts and track the transaction from reorder to delivery. This module will track vendors, their performance, the expediting of spare parts, and administering of any blanket purchasing contracts. In some systems, the purchasing system will also track all service contracts.

Figure I-4 highlights the relationship among the inventory, purchasing, and maintenance systems. As Figure I-4 shows, the purchasing system feeds information (such as costs and vendors) to the inventory system, which ultimately feeds this data into the work order system. A work order is always written against a piece of equipment or a building-floor-room locator. The material costs and history information is passed from purchasing to inventory to the work order and are finally deposited in the correct piece of equipment's history file. This allows for total cost tracking across the entire life cycle of the piece of equipment.

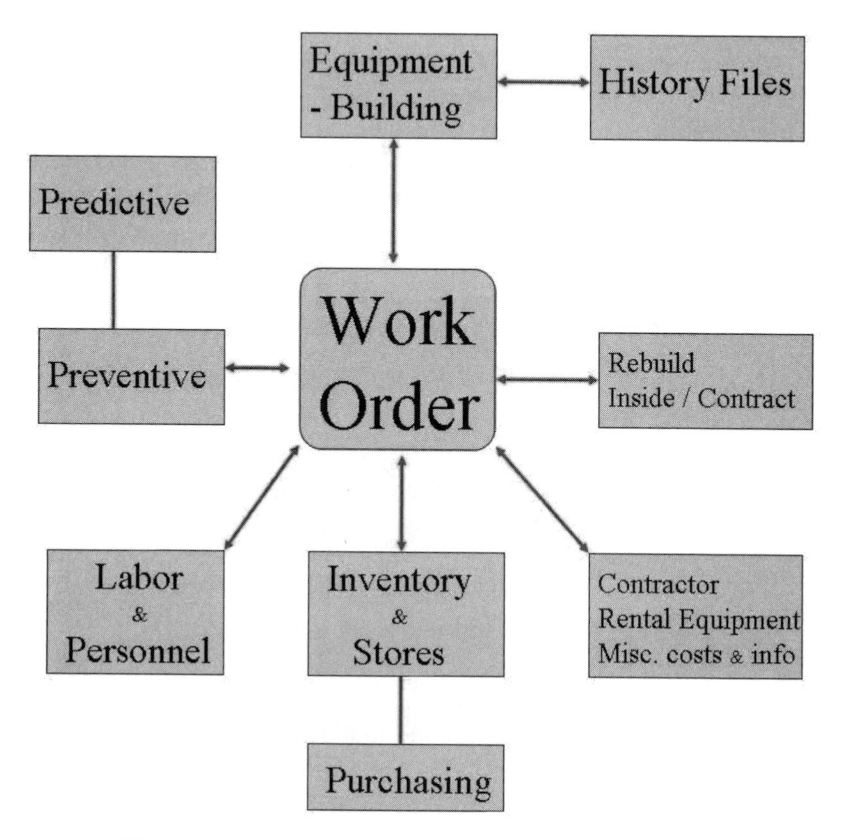

Figure I-4 Work Order and MRO Cost Relationships

If the inventory and purchasing systems are not integrated with the work order system, then collecting accurate cost and history information for all spare parts for the equipment becomes difficult. A fully-integrated system allows for accurate cost and history tracking. Utilizing this type of system should be the goal of every company.

What are the benefits of transitioning from a poorly-managed inventory and purchasing system to one that is considered "best in class?" In a study conducted by Industry Week magazine, a cross-section of companies that had improved their inventory and purchasing systems as part of a computerization effort were tracked. The results showed a significant reduction in total inventory costs in two categories. The first was a 19.4%, lower annual material costs. This reduction was due to proper tracking of materials, avoiding redundant stocking policies, and identifying items that were on hand that could be utilized instead of ordering new parts. The second category was a 17.8% reduction in total inventory levels. When companies computerized their inventory systems, they were able to find duplicate parts stocked in multiple locations. Once they consolidated the inventory, they were able to reduce the total inventory levels and improve delivery service.

Considering these cost reductions, a company holding a $10 million inventory, and having a $10 million annual inventory costs, could

MRO Inventory Best Practices

- 95 - 97% Service Levels
- 100% accuracy of data
- > 1 turn per year on inventory value
- Elimination of non-moving spares
- Reduction of slow moving spares
- Controlled access
- Consignment arrangements
- Strategic partnering with suppliers

Figure I-5 MRO Inventory Best Practices

quickly save over $2 million. Because these would be classified as expense dollars, the savings would drop directly to the bottom line. The results of this survey should help any company quickly see the importance of controlling their MRO inventory and purchasing practices.

What are considered some MRO inventory and purchasing best practices? What are companies that are considered best in class striving to achieve? Figure I-5 highlights some of the typical goals that best-in-class companies are striving towards. For example, best-in-class companies strive to reach 95–97% service levels from the storeroom. This means that 95–97% of the time someone needs a part from the storeroom, the part is available. In addition, best-in-class companies strive to achieve 100% accuracy of data. This means that information such as on-hand quantities, delivery times, and material types are all accurate.

These same companies are focusing on eliminating spare parts that no longer add value, such as nonmoving spare parts. They will also be striving to reduce the numbers of slow-moving spare parts. This they may do by setting up consignment arrangements or strategic partnering with their suppliers. The consignment arrangements and strategic partnering allow some flexibility in the quantities of spare parts that a company must have in their storerooms. Instead of the company stocking these items, many vendors are willing, for a guaranteed amount of business, to keep these spare parts on their stock shelves.

One final point to note on Figure I-5 is the controlled access to storerooms. Companies considered to be best in class always have locked and properly staffed storerooms. This insures that transactions will not occur without being properly recorded and that all stores processes are properly followed. As will be shown in subsequent chapters in the text, this is a very important component to having a successful MRO inventory and purchasing process.

It is the goal of this textbook to present MRO inventory and purchasing processes that will help any company achieve the best in class inventory and purchasing process for their respective business.

OVERVIEW

The Maintenance Strategy Series Process Flow

Good, sound, functional maintenance practices are essential for effective maintenance / asset management strategies. But what exactly are "good, sound, functional maintenance practices?" The materials contained in this overview (and the overview for each of the volumes in the Maintenance Strategy Series) explain each block of the Maintenance Strategy Series Process Flow. They are designed to highlight the steps necessary to develop a complete maintenance / asset management strategy for your plant or facility. The activities described in the Process Flow are designed to serve as a guide for strategic planning discussions. The flow diagram for the Maintenance Strategy Series Process Flow can be found at the end of this overview.

Author's Note

Many individuals may believe that this type of maintenance strategy program is too expensive or time consuming to implement, especially when there are advanced predictive or reliability techniques that might be employed. Yet there is a reason for the sequencing of the Maintenance Strategy Series process flow. If attempts are made to deploy advanced techniques before the organization is mature enough to properly understand and utilize them (basically, the "I want results now" short-term focus), they will fail. The reason? Developing and implementing a sustainable maintenance / asset management strategy is more than just distributing a flow chart or dictionary of technical terms. It is an educational exercise that must change a company culture. The educational process that occurs during a structured implementation of basic maintenance processes must evolve into more sophisticated and advanced processes as the organization develops the understanding and skills necessary.

If an individual is to obtain a college degree, it may involve an investment of four or more years to achieve this goal. Likewise, if a company is to obtain an advanced standing in a maintenance / asset management strategy, it may take up to four years. It is not that someone cannot, through years of experience and education, design their maintenance / asset management process in a short time period. It will, however, take the entire organization (from senior executive to shop floor employees) this amount of time to become mature in their understanding and utilization of the process. Although there will be incremental benefits achieved along the journey to maintenance / asset management excellence, the true benefits are not realized until there is a complete organizational focus on maximizing all aspects of the investment in the assets. It is this competitive focus that separates long-term, sustainable success from a short-term "flash" of improvement.

In the beginning, it is necessary for a plant or facility to decide it is necessary to improve their maintenance / asset management strategy. The business reason for the needed improvement can be multi-faceted, but would likely include:

- Poor Return On Investment (ROI) for the total plant or facility valuation
- Poor throughput for the design of the plant
- Inability to meet production demands
- High cost of occupancy for a facility
- Excessive downtime
- Production inefficiencies

Once the decision has been made to develop / improve the maintenance/ asset management strategy, the Maintenance Strategy Series process flow diagram should be followed. It begins with Preventive Maintenance.

1. Does a PM Program Exist?

Preventive maintenance is the core of any equipment/asset maintenance process improvement strategy. All plant and facility equipment, including special back-up or redundant equipment, must be covered by a complete, cost-effective, preventive maintenance program. The preventive maintenance program must be designed to eliminate all unplanned equipment failures. The preventive maintenance program should be designed to insure proper coverage of the critical equipment of the plant or facility. The program should include a good cross section of the following:

- Inspections
- Adjustments
- Lubrication
- Proactive replacements of worn components

The goal of the program is to insure there will be no unplanned equipment downtime.

2. Is the PM Program Effective?

The effectiveness of the preventive maintenance program is determined by the level of unplanned equipment maintenance that is performed. Unplanned equipment maintenance is defined as any maintenance activity that is performed with less than one week of advanced planning. Unplanned equipment maintenance is commonly referred to as reactive maintenance. An effective preventive maintenance program will reduce the amount of unplanned work to less than 80% of the total manpower expended for all equipment maintenance activities. If more time than this is being spent on unplanned activities, then a reevaluation of the preventive maintenance program is required. It will take more resources and additional time to make progress in any of the following maintenance process areas unless the preventive maintenance program is effective enough for the equipment maintenance to meet the 80%/20% rule.

It should be the goal of not progressing any further until the preventive maintenance program is successful. In addition to requiring more resources and taking longer to develop the subsequent maintenance processes, it is very common to see companies try to compensate for a reactive organization. This means they will circumvent some of the "best practices' in the subsequent processes to make them work in a reactive environment. All this will do is reinforce negative behavior and sub-optimize the effectiveness of the subsequent processes.

3. Do MRO Processes Exist?

After the preventive maintenance program is effective, the equipment spares, stores, and purchasing systems must be analyzed. The equipment spares and stores should be organized, with all of the spares identified and tagged, stored in an identified location, with accurate on-hand and usage data. The purchasing system must allow for procurement of all necessary spare parts to meet the maintenance schedules. All data necessary to track the cost and usage of all spare parts must be complete and accurate.

4. Are the MRO Processes Effective?

The benchmark for an effective maintenance / asset management MRO process is service level. Simply defined, the service level measures what percent of the time a part is in stock when it is needed. The spare parts must be on hand at least 95%– 97% of the time for the stores and purchasing systems to support the maintenance planning and scheduling functions.

Again, unless maintenance activities are proactive (less than 20% unplanned weekly), it will be impossible for the stores and purchasing groups to be cost effective in meeting equipment maintenance spare parts demands. They will either fall below the 95%–97% service level, or they will be forced to carry excess inventory to meet the desired service level.

The MRO process must be effective for the next steps in the strategy development. If the MRO data required to support the maintenance work management process is not developed, the maintenance spare parts costs will never be accurate to an equipment level. The need for this level of data accuracy will be explained in Sections 6 and 10 of the preface.

5. Does a Work Management Process Exist?

The work management system is designed to track all equipment maintenance activities. The activities can be anything from inspections and adjustments to major overhauls. Any maintenance that is performed without being recorded in the work order system is lost. Lost or unrecorded data makes it impossible to perform any analysis of equipment problems. All activities performed on equipment must be recorded to a work order by the responsible individual. This highlights the point that maintenance, operations, and engineering will be extremely involved in utilizing work orders.

Beyond just having a work order, the process of using a work order system needs detailed. A comprehensive work management process should include details on the following:

- How to request work
- How to prioritize work
- How to plan work
- How to schedule work
- How to execute work
- How to record work details
- How to process follow up work
- How to analyze historical work details

6. Is the Work Management Process Effective?

This question should be answered by performing an evaluation of the equipment maintenance data. The evaluation may be as simple as answering the following questions:

- How complete is the data?
- How accurate is the data?

- How timely is the data?
- How usable is the data?

If the data is not complete, it will be impossible to perform any meaningful analysis of the equipment historical and current condition. If the data is not accurate, it will be impossible to correctly identify the root cause of any equipment problems. If the data is not timely, then it will be impossible to correct equipment problems before they cause equipment failures. If the data is not usable, it will be impossible to format it in a manner that allows for any meaningful analysis. Unless the work order system provides data that passes this evaluation, it is impossible to make further progress.

7. Is Planning and Scheduling Utilized?

This review examines the policies and practices for equipment maintenance planning and scheduling. Although this is a subset of the work management process, it needs a separate evaluation. The goal of planning and scheduling is to optimize any resources expended on equipment maintenance activities, while minimizing the interruption the activities have on the production schedule. A common term used in many organizations is "wrench time." This refers to the time the craft technicians have their hands on tools and are actually performing work; as opposed to being delayed or waiting to work. The average reactive organization may have a wrench time of only 20%, whereas a proactive, planned, and scheduled organization may be as high as 60% or even a little more.

The ultimate goal of planning and scheduling is to insure that all equipment maintenance activities occur like a pit stop in a NASCAR race. This insures optimum equipment uptime, with quality equipment maintenance activities being performed. Planning and scheduling pulls together all of the activities, (maintenance, operations, and engineering) and focuses them on obtaining maximum (quality) results in a minimum amount of time.

8. Is Planning and Scheduling Effective?

Although this question is similar to #6, the focus is on the efficiency and effectiveness of the activities performed in the 80% planned mode. An efficient planning and scheduling program will insure maximum productivity of the employees performing any equipment maintenance activities. Delays, such as waiting on or looking for parts, waiting on or looking for rental equipment, waiting on or looking for the equipment to be shut down, waiting on or looking for drawings, waiting on or looking for tools, will all be eliminated.

If these delays are not eliminated through planning and scheduling, then it will be impossible to optimize equipment utilization. It will be the same as a NASCAR pit crew taking too long to do a pit stop; the race is lost by not keeping the car on the track. The equipment utilization is lost by not properly keeping the equipment in service.

9. Is a CMMS / EAM System Utilized?

By this point in the Maintenance Strategy Process development, a considerable volume of data is being generated and tracked. Ultimately, the data becomes difficult to manage using manual methods. It may be necessary to computerize the work order system. If the workforce is burdened with excessive paper work and is accumulating file cabinets of equipment data that no one has time to look at, it is best to computerize the maintenance / asset management system. The systems that are used for managing the maintenance /asset management process are commonly referred to by acronyms such as CMMS (Computerized Maintenance Management Systems) or EAM (Enterprise Asset Management) systems. (The difference between the two types of systems will be thoroughly covered in Volume Four.)

The CMMS/ EAM System should be meeting the equipment management information requirements of the organization. Some of the requirements include:

- Complete tracking of all repairs and service
- The ability to develop reports, for example:
- Top ten equipment problems
  - Most costly equipment to maintain
  - Percent reactive vs. proactive maintenance
  - Cost tracking of all parts and costs

If the CMMS/EAM system does not produce this level of data, then it needs to be re-evaluated and a new one may need to be implemented.

10. Is the CMMS/ EAM System Utilization Effective?

The re-evaluation of the CMMS / EAM system may also highlight areas of weakness in the utilization of the system. This should allow for the specification of new work management process steps that will correct the problems and allow for good equipment data to be collected. Several questions for consideration include:

- Is the data we are collecting complete and accurate?
- Is the data collection effort burdening the work force?
- Do we need to change the methods we use to manage the data?

Once problems are corrected and the CMMS / EAM system is being properly utilized, then constant monitoring for problems and solutions must be put into effect.

The CMMS / EAM system is a computerized version of a manual system. There are currently over 200 commercially produced CMMS / EAM systems in the North American market. Finding the correct one may take some time, but through the use of lists, surveys, and "word of mouth," it should take no more than three to a maximum of six months for any organization to select their CMMS / EAM system. When the right CMMS / EAM system is selected, it then must be implemented. CMMS / EAM system implementation may take from three months (smaller organizations) to as long as 18 months (large organizations) to implement. Companies can spend much time and energy around the issue of CMMS selection and implementation. It must be remembered that the CMMS / EAM system is only a tool to be used in the improvement process; it is not the goal of the process. Losing sight of this fact can curtail the effectiveness of any organization's path to continuous improvement.

If the correct CMMS / EAM system is being utilized, then it makes the equipment data collection faster and easier. It should also make the analysis of the data faster and easier. The CMMS / EAM system should assist in enforcing "World Class" maintenance disciplines, such as planning and scheduling and effective stores controls. The CMMS / EAM system should provide the employees with usable data with which to make equipment management decisions. If the CMMS / EAM system is not improving these efforts, then the effective usage of the CMMS / EAM system needs to be evaluated. Some of the problems encountered with CMMS / EAM systems include:

- Failure to fully implement the CMMS
- Incomplete utilization of the CMMS
- Inaccurate data input into the CMMS
- Failure to use the data once it is in the CMMS

11. Do Maintenance Skills Training Programs Exist?

This question examines the maintenance skills training initiatives in the company. This is a critical item for any future steps because the maintenance organization is typically charged with providing training for any operations personnel that will be involved in future activities. Companies need to have an ongoing maintenance skills training program because technology changes quickly. With newer equipment (or even components) coming into plants almost daily, the skills of a maintenance workforce can be quickly dated. Some sources estimate that up to 80% of existing maintenance skills can be outdated within five years. The skills training program can utilize many resources, such as vocational schools, community colleges, or even vendor training. However, to be effective, the skills training program needs to focus on the needs of individual employees, and their needs should be tracked and validated.

12. Are the Maintenance Training Programs Effective?

This evaluation point focuses on the results of the skills training program. It deals with issues such as:

- Is there maintenance rework due to the technicians not having the skills necessary to perform the work correctly the first time?
- Is there ongoing evaluation of the employees skills versus the new technology or new equipment they are being asked to maintain or improve?
- Is there work being held back from certain employees because a manager or supervisor questions their ability to complete the work in a timely or quality manner?

If these questions uncover some weaknesses in the workforce, then it quickly shows that the maintenance skills training program is not effective. If this is the case, then a duty-task-needs analysis will highlight the content weaknesses in the current maintenance skills training program and provide areas for improvement to increase the versatility and utilization of the maintenance technicians.

13. Are Operators Involved in Maintenance Activities?

As the organization continues to make progress in the maintenance disciplines, it is time to investigate whether operator involvement is possible in some of the equipment management activities. There are many issues that need to be explored, from the types of equipment being operated, the operators-to-equipment ratios, and the skill levels of the operators, to contractual issues with the employees' union. In most cases, some level of activity is found in which the operators can be involved within their areas. If there are no obvious activities for operator involvement, then a re-evaluation of the activities will be necessary.

The activities the operators may be involved in may be basic or complex. It is partially determined by their current operational job requirements. Some of the more common tasks for operators to be involved in include, but are not necessarily limited to:

a. **Equipment Cleaning:** This may be simply wiping off their equipment when starting it up or shutting it down.

b. **Equipment Inspecting:** This may range from a visual inspection while wiping down their equipment to a maintenance inspections checklist utilized while making operational checks.

c. **Initiating Work Requests:** Operators may make out work requests for any problems (either current or developing) on their equipment. They would then pass these requests on to maintenance for entry into the work order system. Some operators will directly input work requests into a CMMS.

d. **Equipment Servicing:** This may range from simple running adjustments to lubrication of the equipment.

e. **Visual Systems:** Operators may use visual control techniques to inspect and to make it easier to determine the condition of their equipment.

Whatever the level of operator involvement, it should contribute to the improvement of the equipment effectiveness.

14. Are Operator-Performed Maintenance Tasks Effective?

Once the activities the operators are to be involved with has been determined, their skills to perform these activities need to be examined. The operators should be properly trained to perform any assigned tasks. The training should be developed in a written and visual format. Copies of the training materials should be used when the operators are trained and a copy of the materials given to the operators for their future reference.

This will contribute to the commonality required for operators to be effective while performing these tasks. It should also be noted that certain regulatory organizations require documented and certified training for all employees (Lock Out Tag Out is an example).

Once the operators are trained and certified, they can begin performing their newly-assigned tasks. It is important for the operators to be coached for a short time to insure they have the full understanding of the hows and whys of the new tasks. Some companies have made this coaching effective by having the maintenance personnel assist with it. This allows for operators to receive background knowledge that they may not have gotten during the training.

15. Are Predictive Techniques Utilized?

Once the operators have begun performing some of their new tasks, some maintenance resources should be available for other activities. One area that should be explored is predictive maintenance. Some fundamental predictive maintenance techniques include:

- Vibration Analysis
- Oil Analysis
- Thermography
- Sonics

Plant equipment should be examined to see if any of these techniques will help reduce downtime and improve its service. Predictive technologies should not be utilized because they are technically advanced, but only when they contribute to improving the equipment effectiveness. The correct technology should be used to trend or solve the equipment problems encountered.

16. Are the PDM Tasks Effective?

If the proper PDM tools and techniques are used, there should be a decrease in the downtime of the equipment. Because the PDM program will find equipment wear before the manual PM techniques, the planning and scheduling of maintenance activities should also increase. In addition, some of the PM tasks that are currently being performed at the wrong interval should also be able to be adjusted. This will have a positive impact on the cost of the PM program. The increased efficiency of the maintenance workforce and the equipment should allow additional time to focus on advanced reliability techniques.

17. Are Reliability Techniques Being Utilized?

Reliability Engineering is a broad term that includes many engineering tools and techniques. Some common tools and techniques include:

a. **Life Cycle Costing:** This technique allows companies to know the cost of their equipment from when it was designed to the time of disposal.
b. **R.C.M.:** Reliability Centered Maintenance is used to track the types of maintenance activities performed equipment to insure they are correct activities to be performed.

c. **F.E.M.A.:** Failure and Effects Mode Analysis examines the way the equipment is operated and any failures incurred during the operation to find methods of eliminating or reducing the numbers of failures in the future.

d. **Early Equipment Management and Design:** This technique takes information on equipment and feeds it back into the design process to insure any new equipment is designed for maintain ability and operability.

Using these and other reliability engineering techniques improve equipment performance and reliability to insure competitiveness.

18. Are the Reliability Techniques Effective?

The proper utilization of reliability techniques will focus on eliminating repetitive failures on the equipment. While some reliability programs will also increase the efficiency of the equipment, this is usually the focus of TPM/OEE techniques. The elimination of the repetitive failures will increase the availability of the equipment. The effectiveness of the reliability techniques are measured by maximizing the uptime of the equipment.

19. Are TPM/ OEE Methodologies Being Utilized?

Are the TPM/OEE methodologies being utilized throughout the company? If they are not, then the TPM/OEE program needs to be examined for application in the company's overall strategy. If a TPM/OEE process exists, then it should be evaluated for gaps in performance or deficiencies in existing parts of the process. Once weaknesses are found, then steps should be taken to correct or improve these areas. Once the weaknesses are corrected and the goals are being achieved, then the utilization of the OEE for all equipment relate decisions is examined.

20. Is OEE Being Effectively Utilized?

The Overall Equipment Effectiveness provides a holistic look at how the equipment is utilized. If the OEE is too low, it indicates that the equipment is not performing properly and maximizing the return on investment in the equipment. Also, the upper limit for the OEE also needs to be understood. If a company were to focus on achieving the maximum OEE number, they may pay too much to ever recover the investment. If the OEE is not clearly understood, then additional training in this area must be provided. Once the OEE is clearly understood, then the focus can be switched to achieving the financial balance required to maximize a company's return on assets (ROA).

21. Does Total Cost Management Exist?

Once the equipment is correctly engineered, the next step is to understand how the equipment or process impacts the financial aspects of the company's business. Financial optimization considers all costs impacted when equipment decisions are made. For example, when calculating the timing to perform a preventive maintenance task, is the cost of lost production or downtime considered? Are wasted energy costs considered when cleaning heat exchangers or coolers? In this step, the equipment data collected by the company is examined in the context of the financial impact it has on the company's profitability. If the data exists and the information systems are in place to continue to collect the data, then financial optimization should be utilized. With this tool, equipment teams will be able to financially manage their equipment and processes.

22. Is Total Cost Management Utilized?

While financial optimization is not a new technique, most companies do not properly utilize it because they do not have the data necessary to make the technique effective. Some of the data required includes:

- MTBF (Mean Time Between Failure) for the equipment
- MTTR (Mean Time To Repair) for the equipment
- Downtime or lost production costs per hour
- A Pareto of the failure causes for the equipment
- Initial cost of the equipment

- Replacement costs for the equipment
- A complete and accurate work order history for the equipment

Without this data, financial optimization can not be properly conducted on equipment. Without the information systems in place to collect this data, a company will never have the accurate data necessary to perform financial optimization.

23. Are Continuous Improvement Techniques Utilized for Maintenance / Asset Management Decisions?

Once a certain level of proficiency is achieved in maintenance/ asset management, companies can begin to lose focus on their improvement efforts. They may even become complacent in their improvement efforts. However, there are excellent Continuous Improvement (CI) tools for examining even small problems. If new tools are constantly examined and applied to existing processes, all opportunities for improvement will be clearly identified and prioritized.

24. Are the CI Tools and Techniques Effective?

This question may appear to be subjective; however, improvements at this phase of maturity for a maintenance / reliability effort may be small and difficult to identify. However, the organizational culture of always looking for areas to improve is a true measure of the effectiveness of this step. As long as even small improvements in maintenance / reliability management are realized, this question should be answered "Yes."

25. Is Continuous Improvement Sought After in All Aspects of Maintenance / Asset Management?

When organizations reaches this stage, it will be clear that they are leaders in maintenance / reliability practices. Now, they will need continual focus on small areas of improvement. Continuous improvement means never getting complacent. It is the constant self-examination with the focus on how to become the best in the world at the company's business. Remember:

Yesterday's Excellence
is
Today's Standard
and
Tomorrow's Mediocrity

Maintenance Strategy Series Part 1

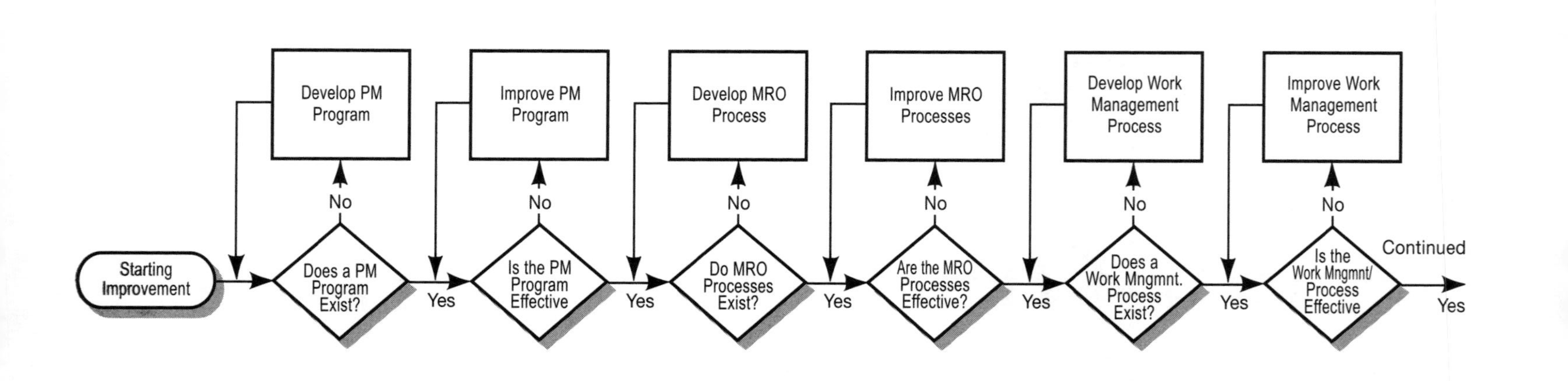

Maintenance Strategy Series Part 2

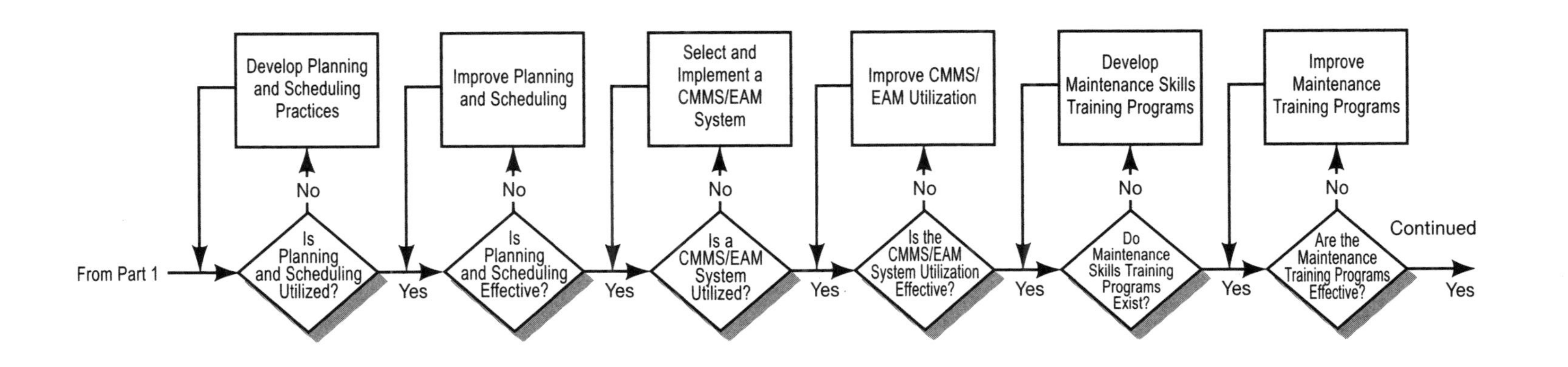

Maintenance Strategy Series Part 3

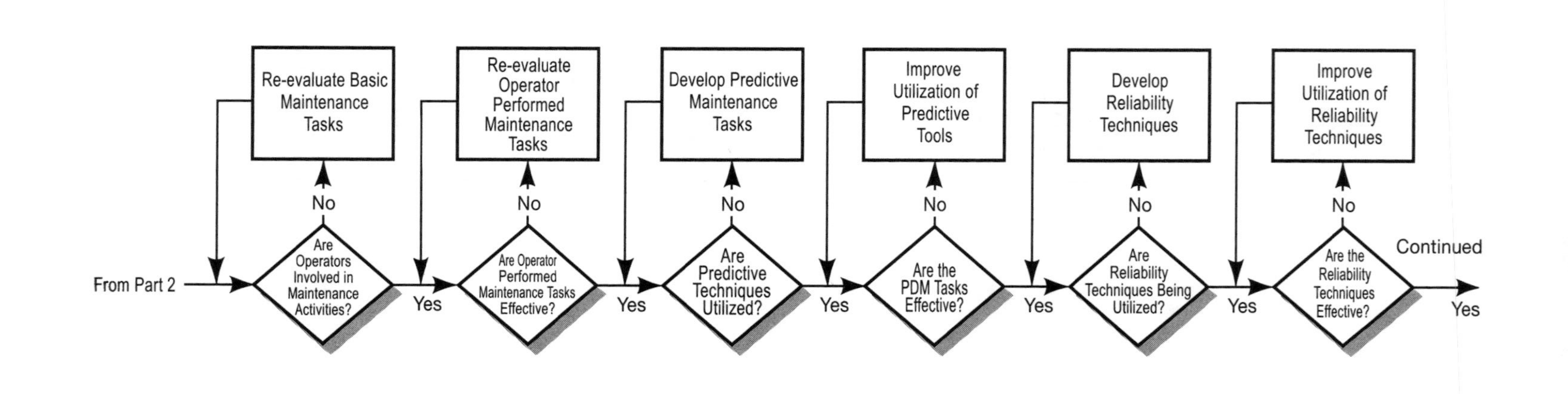

Maintenance Strategy Series Part 4

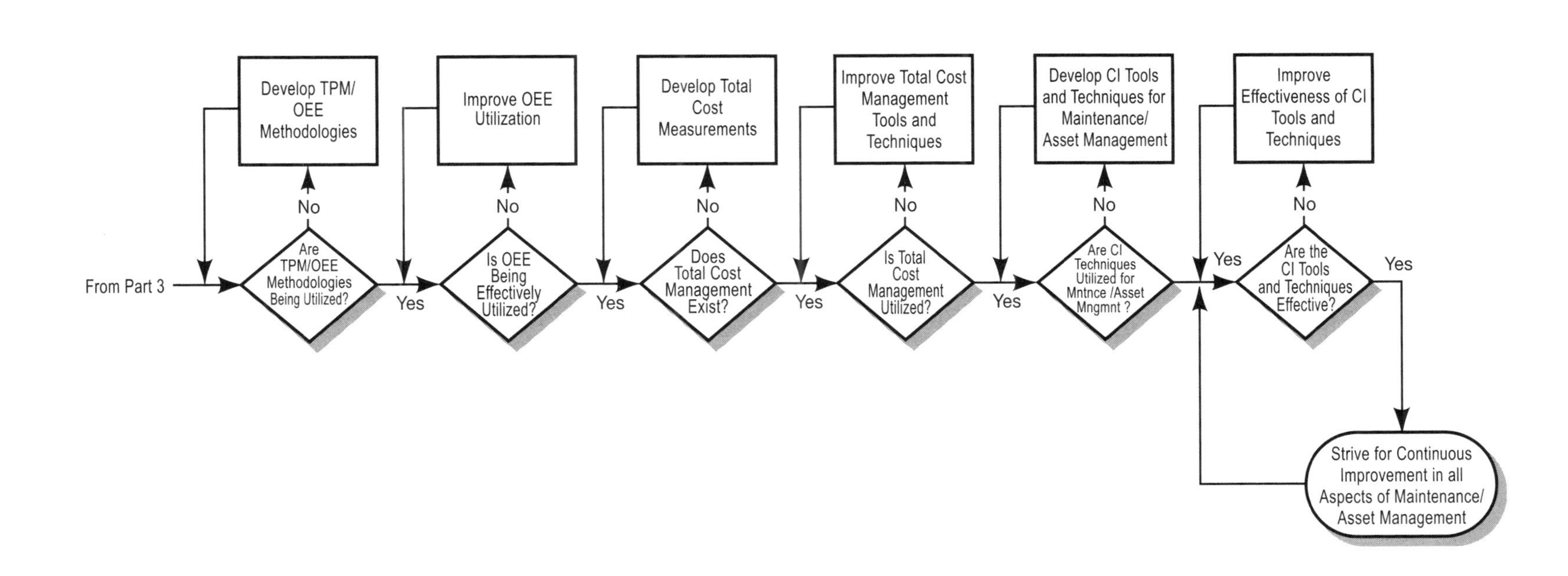

1

Organizing MRO Inventory and Purchasing

MRO Analysis

MRO inventory and purchasing, in this text, stands for maintenance, repair, and overhaul inventory and purchasing. It is the organizational function responsible for procuring and storing spare parts for the assets or equipment in the plant or facility. However, the impact of this organization is pervasive. Its efficiency affects all of the departments (such as maintenance, project engineering, contractors, etc.) that perform work on the company's assets or equipment.

MRO inventory and purchasing are the single biggest maintenance support functions that contribute to low maintenance productivity. As was covered in the introduction to this text, many maintenance labor delays are related to issues concerning materials and spare parts. Because the purchasing function typically procures the materials for the MRO storerooms, it also has a big impact on maintenance productivity. Additional material problems that impact labor productivity include late deliveries from vendors, wrong parts delivered, and parts that are damaged and cannot be used when they arrive. If you believe your organization has problems in the MRO inventory and purchasing areas, you should conduct an internal analysis to understand clearly the type and severity of the problem.

The following questions can be used for a high-level analysis of the MRO inventory and purchasing systems in most companies. For problems that are highlighted by using this high-level analysis, the information detailed later in this text should help managers make the proper changes in their processes to bring the MRO inventory and purchasing functions up to an acceptable level.

The first 10 analysis points are:

1. Do you have detailed MRO procedures covering:
    - The material/ spare part ordering process?
    - The expediting process?
    - The delivery acceptance process?
    - The receipt and binning process?
    - The spare parts issuing process?

2. Do the maintenance and purchasing staff consistently follow these procedures?

3. Are stock replenishments governed by established formulas and formal inventory rules?

4. Do you and your organization know what your total ordering cost is?

5. Do you and your organization know what your inventory holding cost is?

6. Does your organization have an effective salvage and reclamation program?

7. Does your organization have an ongoing obsolescence program?

8. Is finding local resourcing an ongoing effort?

9. Is the service level adequate? (95–97%)

10. Is access to inventory information easily available to all personnel at all times?

Are your costs consolidated clearly?
- Can you easily differentiate between spare parts charges and contract labor and material charges?
- Are all spare parts charges debited to specific equipment items (not department charge codes)?

Do you have localized accounting practices?

- Does one area use a $100 limit to differentiate between an inventory vs. expense item whereas another area uses $1000?

Are your material classifications clearly defined?

- Can the maintenance, storeroom, and purchasing personnel differentiate between spares parts and consumables?
- Can the maintenance, storeroom, and purchasing personnel differentiate between operating supplies and maintenance materials?

Does your cost tracking stopping at the invoiced amount level?

- Do the maintenance, storeroom, and purchasing personnel understand the true cost to procure the materials?
- Do the maintenance, storeroom, and purchasing personnel understand the true cost to hold the materials?

If these questions highlight some deficiencies in the MRO storeroom operation, or if a company is just instituting an MRO storeroom, reviewing the basics of the inventory and purchasing process will be helpful.

A flow diagram that can be utilized to review basics of the MRO stores and purchasing process is found in Figure 1-1. The items on this flowchart can be used to develop the MRO basics. They may also be considered prerequisites to developing an MRO storeroom and purchasing process.

The flow begins with the decision to develop an MRO process or to improve an existing MRO process. The second step is to develop MRO storeroom locations and organize the storeroom. The next step then identifies all of the spare parts that are going to be stocked in the storeroom and develops an identification system that allows for quick and easy identification of each of the spare parts.

Once the number parts have been determined, then it is necessary to develop the stocking policies. These can be based on historical use each or can involve the utilization of some advanced statistical calculations to determine proper stocking policies. Once the volume of materials is clearly understood, the storage areas and the equipment to move the spare parts in and out of the storage areas can be determined.

MRO Spare Parts Pre-Requisites

Developing MRO Basics

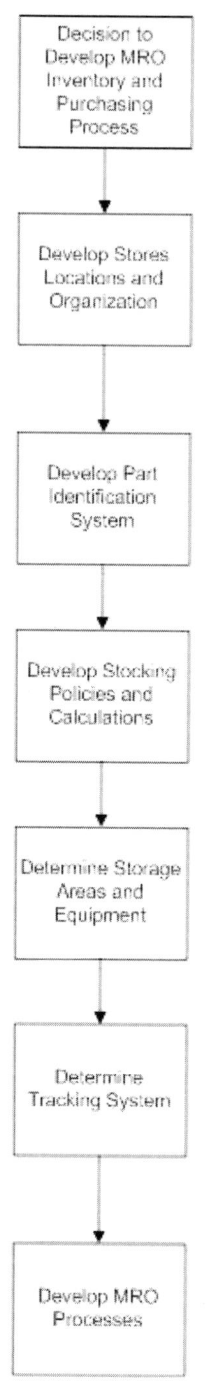

Figure 1-1 Developing MRO Basics

With a sizable inventory, manually tracking the data can become rather cumbersome. For this reason, most organizations have computerized their inventory and purchasing systems. Most MRO inventory and purchasing will use some form of corporate or plant-level system that is already established. In the event that the tracking system is not at the corporate or plant level, a departmental system can be used. However, these systems are usually not as effective because the data is kept at a low departmental level, rather than at the plant or corporate level. This makes cost tracking, discounting, blanket purchasing, and strategic partnering with vendors more difficult.

Organization Configurations

There is a constant conflict between the maintenance organization and the MRO inventory and purchasing organization. The maintenance point of view is highlighted in Figure 1-2. The maintenance concerns are focused on having the part on hand when it is needed. When maintenance is in a proactive mode, this is typically not an issue. However, when maintenance is in a reactive mode, it is virtually impossible for a storeroom to have every part that might be needed on a moment's notice. It is for this reason that maintenance must be moving towards creating a proactive environment. If the maintenance organization is not proactive, it is unlikely that MRO inventory and purchasing will be successful.

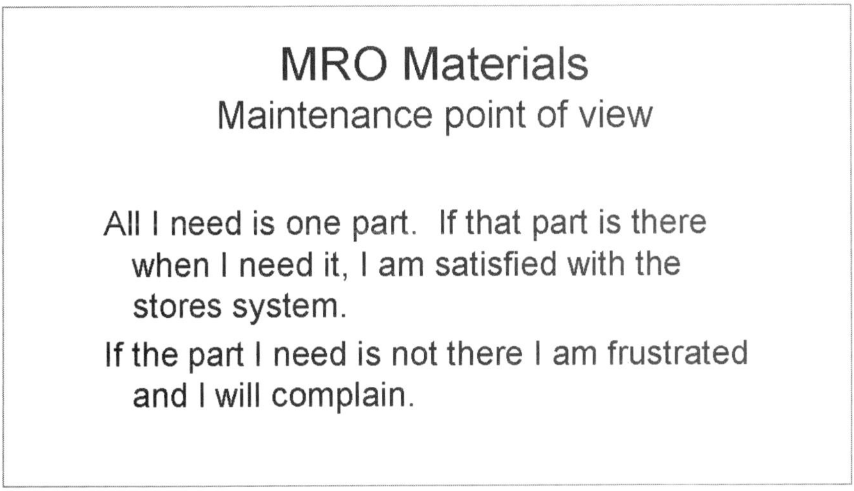

Figure 1-2 MRO Materials—Maintenance

MRO Materials
Inventory point of view

Can I fill the next order promptly?
What is my safety stock?
Have I reached my reorder point?
What is the economic order quantity?
Is usage changing?
Have the sensitivities to cost, time, or support changed?

Figure 1-3 MRO Materials—Inventory

Figure 1-3 highlights the inventory point of view for MRO materials. As can be seen, the MRO organization is interested in providing good service to the maintenance organization. The MRO organization focuses on using good inventory controls to accomplish this goal. So the MRO organization is concerned about their safety stock, the reorder point, the economic order quantity, and usage patterns. The MRO organization will also need good, complete data to track the proper costs, delivery times, and levels of support that are necessary, always focusing on the lowest overall cost.

The conflict between maintenance and the MRO organization arises when neither sees the big picture. If they can focus on being able to deliver equipment capacity, they can be successful. The MRO organization supplies the spare parts; the maintenance organization then uses the spare parts to ensure that the company's assets perform properly. Without focusing on a partnership to accomplish this, there will always be conflicts between the two organizations.

Stores Locations and Organization

When considering stores locations and organizations, one of the primary factors is how the maintenance department is organized. If the maintenance organization is centralized, then it is usually best to have the stores locations centralized as well.

Figure 1-4 shows the advantage of this configuration. This diagram shows that when maintenance and stores are together, in the center of a plant or facility, it is easy to travel to the equipment. The distance that maintenance has to travel to obtain spare parts and then travel to the equipment is minimized. This arrangement maximizes labor productivity for the maintenance department. It also allows for more rapid response when equipment experiences problems. Therefore, for geographically compact plants or facilities, this is an ideal configuration.

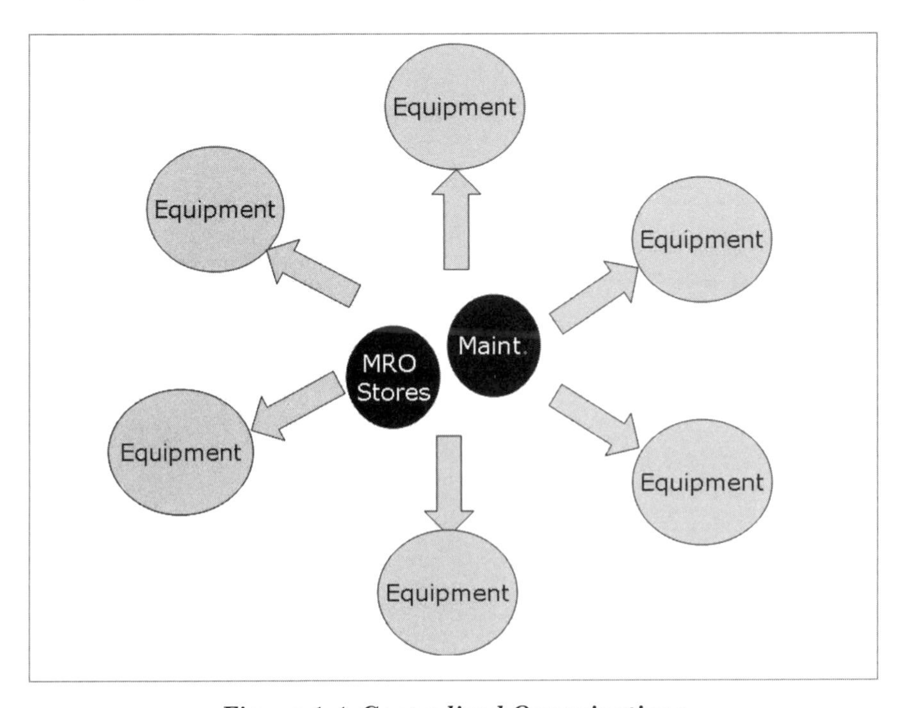

Figure 1-4 Centralized Organizations

The problem that is typically experienced when designing this type of an organization is trying to find the centralized space for both maintenance and the MRO storeroom. Most companies are reluctant to give up prime space for these two departments. However, once the total cost picture is clearly understood, it is a good business decision to centrally locate maintenance and MRO stores for smaller geographically compact plants.

If the maintenance organization has an area configuration, then it is usually best to have the stores locations in an area configuration as well.

Figure 1-5 highlights this configuration. In this figure, there is a larger geographical footprint for the plant or facility. It may actually be several miles from one side of the diagram to the other. Given this condition, it makes sense to have MRO stores and maintenance located at opposite sides of the plant. This minimizes the travel time for maintenance to get to the equipment. Increased maintenance labor productivity is the result, whether it is in a proactive or a reactive environment.

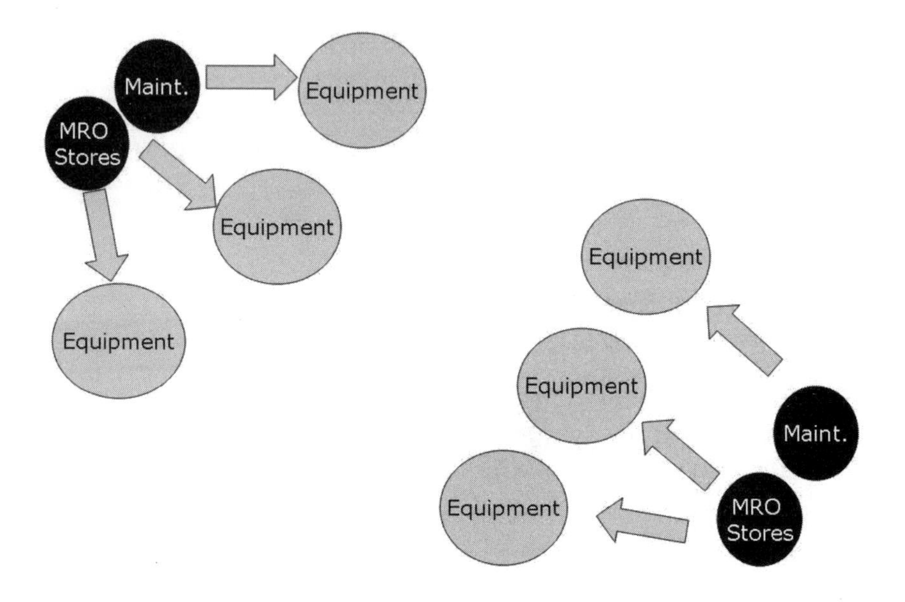

Figure 1-5 Area Organizations

The common complaint or using this type of a structure is the increased stocking levels for the MRO storeroom. In this configuration, it is true that there will be an increase in the stocking levels. This must be balanced against the increased maintenance labor productivity and the reduced downtime that will be achieved by using this configuration.

If the maintenance organization is in a combination configuration, then it is usually best to have the stores locations in a combination configuration. Figure 1-6 highlights a combination configuration. A combination organization is typically used on large plants. Although maintenance and MRO stores maintain an area presence, certain parts of the organization and types of spares are brought to a centralized location.

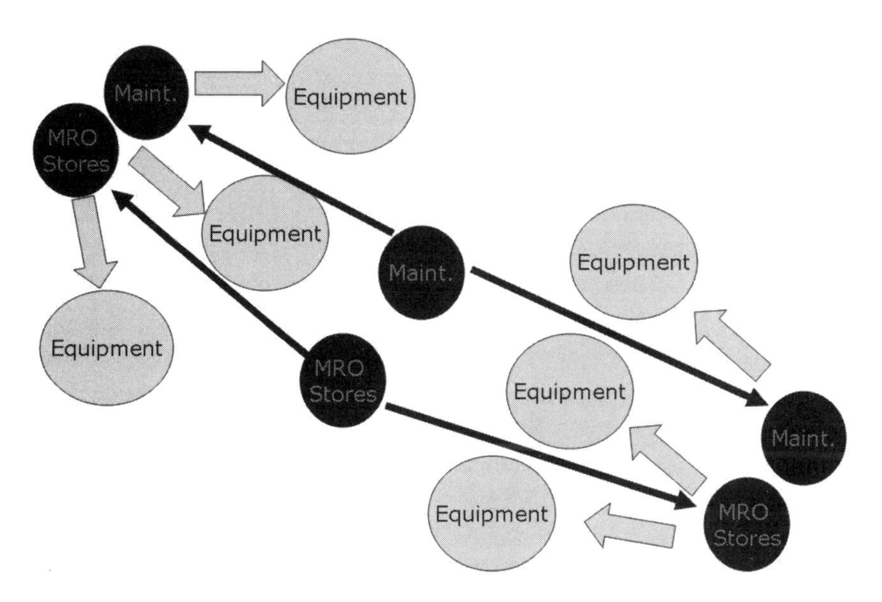

Figure 1-6 Combination Organizations

In a maintenance department utilizing a combination organization, the central trades group will be comprised of technicians who would not be fully utilized in a single area. The combination organization allows the sharing of these central trade technicians among several areas to increase their overall utilization. The same reasoning applies to the MRO stores. There are certain common spare parts that can be kept centrally, while parts specific to a certain area should be kept in the area. This configuration works well for large organizations and optimizes both maintenance resources and MRO spare parts, and also increases equipment availability.

The reason for matching the configuration is the impact that travel time to and from the storeroom can have on maintenance labor productivity. If there is a mis-match between the stores and the maintenance organizations, then the travel time to and from the stores to procure spare parts can be considerable. Figure 1-7 highlights this issue. As can be seen from this figure, the distance (and time to travel) can be considerable if maintenance needs to travel to the equipment. If the maintenance technicians need to go to the storeroom before going to the equipment, they would face even more travel time. If the equipment is down, waiting on maintenance, then the losses increase to include not just lost maintenance productivity, but also lost production. In geographically compact plants, this may not be an issue: however, in large plants, the losses incurred with

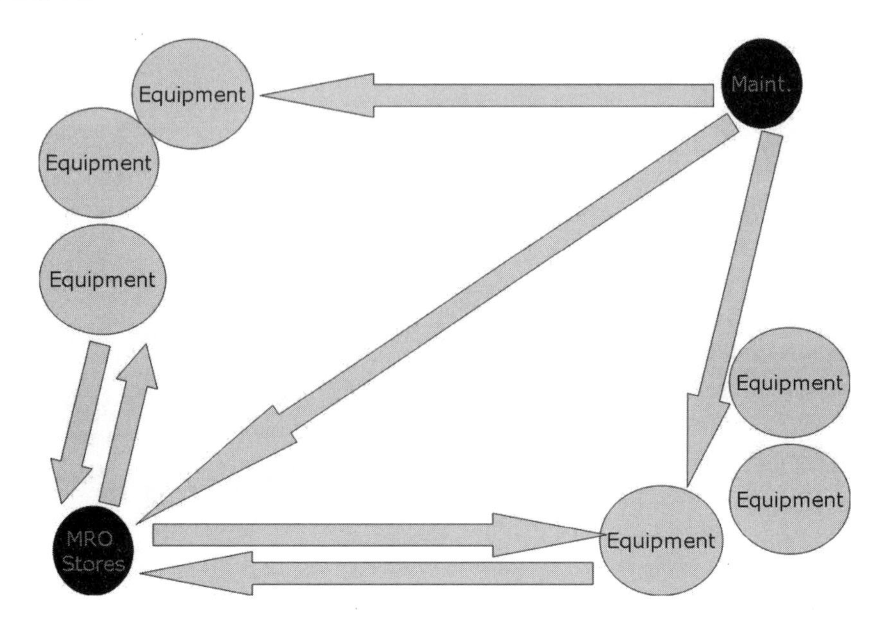

Figure 1-7 Dysfunctional Organizations

dysfunctional organizations can have a dramatic impact on profitability.

This example of lost maintenance productivity and lost production is particularly true in the early stages of developing a maintenance organization because a reactive organization needs spare parts on short notice. As the maintenance organization matures and becomes better at planning and scheduling, material demands can be specified weeks in advance. With this level of forecasting, kitting and delivery systems can be implemented and the location of the storeroom becomes less of an issue.

The decision on the location of the storeroom and possibly the number of locations needs to be carefully balanced. The balance is between lost maintenance productivity and the cost of inflating the inventory to support multiple locations. Some organizations, in an attempt to increase maintenance productivity, will carry excessive numbers of spare parts to compensate for organizational mismatches. Delayed spare parts procurement does not impact only maintenance productivity; it also impacts equipment availability. When making the decision about spare parts locations and organizational structures, it is important to consider all of the factors, then come up with a financially balanced decision.

Part Identification

In addition to the stores organization, it is necessary to develop a part information database for all of the parts that will be stocked in the storeroom. One of the first steps in this process is to clearly identify all of the spare parts. Once they are clearly identified, it is necessary to number each of the spare parts. In some organizations, the part number is merely some type of a sequential number. However, the preferred method is to make an intelligent numbering system for spare parts. Parts should be identified with some type of class or subclass, then the second part of the number should be a specific identifier. Even a description of the part could be included in the number.

In some companies, commodity codes are used to start the part number. This code indicates the type of item that is being specified, and is usually a 2- or 3-digit number. Figure 1-8, which provides a sample numbering system, starts with a type code. This type code will set the general classification, or the spare part. For example, type code 000 will always be the class for mechanical power transmission. Type code 200 will always start the fluid power spare parts. Type code 500 will always be fasteners.

Type Code		Component Code	Description
000		**000**	Mechanical Power Transmission
010		A10	Belts, Flat
010		A20	Belts, V
010		A30	Belts, Timing
010	**B00**		**Chains, General**
010	B10		Chain, Roller
010	B20		Chain, Conveyor
200		**A00**	Fluid Power
200		A10	Pumps, Vane
300		**A00**	Fluid Handling
400		**A00**	Piping
500		**A00**	Fasteners
600		**A00**	Electrical
700		**A00**	Electronic
800		**A00**	Instrumentation
900		**A00**	Facility Related

Figure 1-8 MRO Part Numbering Guide

Type code is 000 and can be subdivided into component codes. For example, 010, a subset of type code 000, would be designated for belts. Component code 810 would be for flat belts. Component code 820 would be for V- belts. Developing the number earlier, a V-belt number would begin with 010-A20-XXX, with the XXX being the specific belt.

Using an intelligent numbering scheme can be very useful in controlling MRO spare parts. Using the vendor's part number or an unintelligent part number can make finding spare parts more difficult. This is particularly true, depending on the type of inventory tracking system that is used. Even though it may take more time to develop an intelligent part numbering system, it has always proven to be beneficial in the long run.

Stocking Policies and Calculations

Numerous methods can be utilized to manage stocking levels and some can become quite complex. For this text, we will focus on four terms and how they can be used to improve the management of the inventory.

Maximum On-Hand Quantity

The first term is the maximum on-hand quantity. This quantity is the maximum number of spare parts that are to be on hand in the storeroom at any given time. While, there are sophisticated mathematical models that can be used to determine this quantity, it is sufficient to note that this number should provide the service level percentage that is cost effective for the company. The service level is the percentage of time that the part will be on hand (in sufficient quantity to meet the demand) when it is required. An appropriate service level is in the 95%+ range.

Minimum On-Hand Quantity

The second term is minimum on-hand quantity. This quantity is the fewest number of spare parts that should be on hand at any given time. For some spare parts, the minimum on-hand quantity is the same as the reorder point. However, this is not always the case. The reorder point must be set at a level that factors in the delivery time for the spare part. If the minimum on-hand quantity is the reorder point, and a demand for the part occurs before the order is placed and received, then the on-hand quantity will drop below the minimum. Then, depending on the minimum levels that are set, a stock-out could possibly occur. Therefore, the re-order point is expressed more correctly as the minimum on-hand quantity plus the safety stock (the stock to cover the order and delivery times).

Reorder Quantity

The third term is the reorder quantity. This is the quantity that will be specified when an order is placed with the vendor. It will typically be the difference between the maximum on-hand quantity and the minimum on-hand quantity. However, there are some usage patterns that will require a different reorder quantity. Again, these can be determined by using some mathematical algorithms.

Reserve Quantity

A fourth term is the reserve quantity. The reserve quantity is the number of spare parts on hand that have been planned and scheduled for a particular job or series of jobs. Technically, the reserved quantity should be subtracted from the on-hand quantity to see if the minimum on-hand quantity, or reorder point, is reached. Because these are items that should be consumed in the next two-to-four weeks, a reorder should be triggered. Using a reserve quantity should ensure that there are no surprise demands that would allow a stock out to occur.

Inflated Inventories

The following scenario could cause inflated inventories:

- A planner is planning work and having it placed in the backlog. As materials are planned for the work, the reserved items are driving the on-hand quantity below the minimum on-hand quantity or reorder point. The system will see the need to re-order the spare part and order the difference between the maximum and minimum quantities. As the work that was planned is carried out, some materials are not used. These items are returned to the storeroom. The on-hand quantity for these items is now above the reorder point, or the minimum on-hand quantity.
- Because an order has already been placed, there is a shipment due. The quantity of the shipment is the difference between the minimum on-hand quantity and the maximum on-hand quantity. Because the on-hand quantity is no longer at the minimum on-hand level, or the reorder point, when the shipment is received, the on-hand quantity will now be over the maximum on-hand

quantity.

- This scenario clearly highlights the importance for planners to be very accurate when planning material demands for work orders. In one particular case, a company, which used a centralized depot for supplying materials to five plants, had over $20 million in spare parts that were simply over-maximum.
- Understanding the relationship between maximum, minimum, reorder quantity, reserved quantities, and reorder point is critical to properly managing any spare part inventory.

Determining Storage Areas and Equipment

Another major prerequisite to properly establishing storeroom and purchasing departments is to define the storage area and the types of storage equipment required. One of the first issues that must be addressed is the footprint of the storeroom area. The area of the storeroom must be a balance between having enough storage for all the spare parts and the cost of the storage area itself. Most companies do not provide sufficient space to store spare parts.

This has led to the development of an industry to provide storage devices to optimize the space that is available. For example, Figure 1-9 shows a shelf and box arrangement. This particular example uses card-

Figure 1-9 Shelving and Box Arrangement

board boxes correctly labeled on the end for the spare parts that are contained in the box. In addition to the cardboard boxes, there are also plastic and — for some specialized spare parts — metal boxes. This shelving and box arrangement allows the spare parts to be stored, assigned a shelf-bin location, and quickly retrieved when required. This arrangement is quite common in most MRO storerooms.

Another storage arrangement that is quite useful is pictured in Figure 1-10. This is known as a stacked drawer arrangement. Each of the drawers has small partitions where individual spare parts can be stored. Each of the small partitions is labeled with the appropriate part number. This arrangement is very useful because each part can be given an aisle, bin, drawer, and partition designation.

In addition, there are many practices that have evolved over the years, for storing special items. For example, Figure 1-11 shows a storage arrangement for various types of belts. This arrangement allows the belts to be clearly visible with the appropriate part number fastened to the wall just above them.

There are many specialty providers for MRO spare parts. When developing an MRO storage area, a company should contact these providers to see what options are available. In many cases, they have been able to provide solutions to storage problems that overcome space and cost requirements.

One final point about storing major spare parts is highlighted in

Figure 1-10 Stacked Drawer Arrangement

Figure 1-11 Wall Display

Figure 1-12 Outside Storage

Figure 1-12. In some cases, companies store major spare parts outside the building, uncovered, and exposed to the environment. This practice is considered an extremely poor one because these spare parts will rapidly deteriorate in the outside environment. The combination of temperature change and relative humidity will cause rust and corrosion of the spare parts. Some companies have even gone to the measure of wrapping the spare parts in plastic. This solution, however, does not compensate for being stored outside. The simple heating and cooling during a day-night cycle will accumulate moisture on the inside of the plastic wrapping. This increase in moisture eventually damages the equipment.

Companies have tens, if not hundreds, of thousands of dollars of equipment in storage. It is a tremendous waste of resources if these spare parts are damaged during their time in storage and need to be scrapped or repaired before they can be used. Good common sense should be used when storing major spare parts.

Determining the Tracking System

When this text refers to a tracking system, it means either a computerized maintenance management system (CMMS) with a stand alone MRO inventory system or an enterprise asset management (EAM) with an integrated corporate MRO inventory system. A typical information and transaction flow for an EAM MRO control system is shown in Figure 1-13.

The core of the program is the store's master data file. The master data file stores much of the spare parts nomenclature. As mentioned previously, it will contain the identifying part number, description, and much of the data concerning stocking levels reorder points and reorder quantities. Five principal transactions occur in the system. These are: part reservations, part issues, purchase order information, receipts, and finally, the transactional detail posting to accounting.

The maintenance module, especially the planning and scheduling component, generates spare part demands from work orders. These orders appear in the system as stock issue requests, which are basically reservations for parts. There can be other part request outside the work order system, such as contractor demands, or project demands. Based on these requests, parts are issued. These show in the system as stock issues.

The information concerning the stock issues is fed back to the maintenance module, where the spare part charges appear against a specific work order number. There can be additional stock issue requests traveling to the maintenance module that do not originate during planning and scheduling. These are typically called a window-issued spare part because the maintenance technician walks to the storeroom window and specifically asks for the spare part. Of course, the maintenance technician must

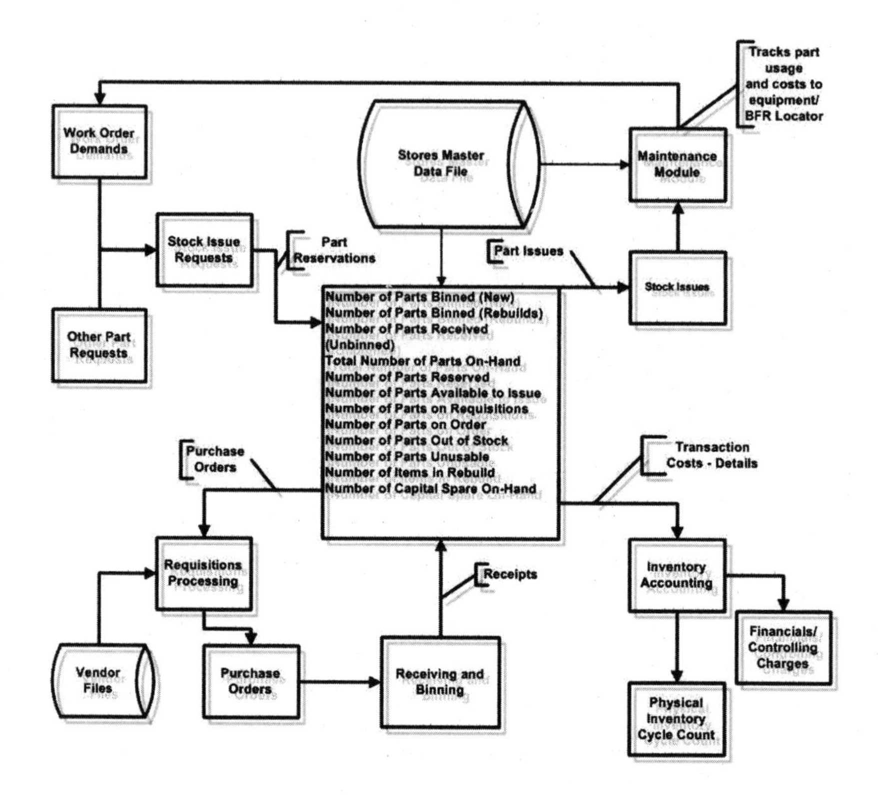

Figure 1-13 MRO Inventory and Purchasing Transaction Flow Diagram

charge the spare part to a work order number. This completes the cost-tracking loop.

As items are used from the storeroom, they will reach their reorder point or minimum on-hand quantity. This then necessitates the initiation of a purchase order. The purchase order will start out as a purchase requisition. It is then processed, a vendor selected, and a purchase order finally issued. The purchase order is sent to the vendor, who processes the order and ships the materials. The materials are then received, and placed in the appropriate storage location or bin. All of the necessary receipt information is completed and filed in the store's master data file.

As materials are used, and then reordered, financial transactions occur. The financial transactions are tracked through inventory accounting. The inventory accounting module must then reconcile the financial transactions against the plant's or corporation's financial and controlling system. In an integrated system, such as SAP, this occurs automatically. In some maintenance and accounting systems that are merely interfaced, these transactions must be updated in a batch mode. This is an extra step that may get overlooked in some maintenance organizations; overlooking it results in poor tracking of the stores financials.

In addition to items issued and tracked from the storeroom, there are some items that vanish from the storeroom. This occurs for various reasons. However, the missing stock must be accounted for financially. Thus inventory cycle counts are performed to ensure stock accuracy. When items are found missing, this value must be debited against the financial system.

These systems are closed loop and completely automated. The maintenance and source personnel only need to enter the required data, and most good systems will properly track all the information.

How do these systems simplify inventory tracking? They must have certain basic features to be effective and tracking MRO spare parts.

First, a good inventory system should track balances for all items, including all part issues, all part reservations, and all return to storage.

Second, the system should maintain a parts listing for each piece of equipment. This listing is commonly referred to as a bill of material or BOM. There is also a reverse of the BOM, called the "where-used" listing. This listing begins with a part number, and then displays all of the equipment where the part can be used as a spare.

The system should also be able to track item repair costs and movement history. This function is particularly important for any organization that uses rebuilt spares. This function allows for the tracking of the spare part, what the repair cost was, and where (at any particular moment in time) the spare part is located (whether it is installed, in storage, or at the repair shop).

The system should also be able to cross-reference any spares to substitute spares that are also in stock. This is important because, if an item is out of stock, it makes it easier to obtain a backup spare part.

The fifth important feature is the ability to place a reservation for a spare part for a job that will be scheduled in the near future. This feature prevents someone from removing the spare part from storage for another job. It also helps to prevent scheduling conflicts.

Another important feature is the system's ability to notify a requester, usually a planner, when items are received for a job. When a part is out of stock, and in order needs to be placed, the system can track the order through to receipt, and then immediately notify the requester that the job can be scheduled. This allows for timely completion of work, waiting on parts.

The seventh feature that is important is the ability to generate work orders to fabricate or repair an item. If a specialized item is required to complete a job, a work order can be generated and attached to the primary work order for the fabrication. In addition, if an item is a rebuildable spare, it must be tracked from repair to return to installation. The ability to generate a work order for this process allows for proper cost tracking.

These systems should also be able to track when an item reorder is needed, with reorder points, or maximum/minimum quantities, and track the order through the purchasing process to receipt. This data allows a planner to track materials, then schedule work based on the projected delivery dates. This feature will keep the maintenance backlog current.

The ninth feature that is essential is the ability to track a requisition, purchase orders, specialized orders, and the receipt of all items. These features are required just to ensure that the system will always have the information needed for the planner to control the workflow process.

Finally, the system should be able to produce performance reports for inventory and purchasing. Common reports would include inventory accuracy, stock turnover, stock outs, service level, and purchase cycle time. These reports will ensure that the inventory and purchasing functions are providing good, timely service for the maintenance department.

Although many inventory systems will have features beyond the basic ten described here, all systems should be checked to ensure these basic items are covered. Without these ten features, managing an MRO inventory and purchasing function will be very difficult.

When the basic foundation discussed in this chapter and shown previously in Figure 1-1 is in place, moving ahead and developing specific MRO processes for the plant or department will be easier.

MRO Definitions and Transactions

Finding Balances in MRO Management

MRO materials management is a decision-making process. It requires balancing financial differences between competing interests.

1. Service Level vs. Stock Out

At a 90% service level, there is a 10% chance that a part will not be available when required. The costs associated with a stock out include lost production (downtime cost), materials expediting cost, reduced maintenance labor productivity, etc.

If the service level is increased to 95%, then there is only a 5% chance that a part is not available. Although this change reduces the probability of incurring a stock-out cost, the additional inventory stocking level increases the capital investment costs, the holding cost, the size (and cost) of the storeroom, etc.

2. Cost of Safety Stock vs. Service Level

The more inventory that a company holds, the higher that its investment cost will be. If a company sets the safety stock at a high level (reorder point = minimum on hand quantity + safety stock), it will incur higher costs. Conversely, a higher investment cost generally corresponds to a higher MRO stores service level.

The question facing a company is "What service level can you afford?"

100%	Capital investment too expensive for almost any plant
95%	A good target for most companies
90%	Downtime cost will be too high for almost any plant

3. Holding Cost vs. Ordering Cost

When larger quantities are ordered, they increase the holding cost, which is based on the total investment in the spare parts. However, ordering costs are lowered, such as paper work, purchasing time, expediting time, transportation, etc. The issue here is to balance the ordering costs with the holding costs.

Issues that Influence the Balance

What are some of the issues that influence the balance between ordering costs and holding costs? The following is a list of the factors.

1. Vendor or manufacturer order fulfillment lead-time
    - How much time from order to delivery?

2. Variability of delivery lead-time
    - Some times a day, some times a week
    - Weather or seasonal-related impacts

3. Service level and safety stock
    - Risk of using up all the stock before the order arrives

4. Variability of usage
    - The plant used 1 in January, 13 in February, 8 in March

5. Other Issues influencing stocking decisions
    - Vendor or manufacturer is prone to labor or strike problems
    - Equipment that the spare parts are used on is no longer manufactured
    - Geographical location of vendor or manufacturer

There are two general rules that balance MRO inventories. They are:

1. The larger the safety stock, the lower the risk of stock out and the higher the cost of holding inventory
2. The smaller the safety stock, the higher the risk of stock out and the higher the cost of purchasing

Determining Proper Stocking Policies

Based on the type of items that are carried in an MRO storeroom, it can be seen that one policy will not work for all inventory items. Some very expensive items are slow-moving, whereas other very inexpensive items move very quickly.

What are some of the factors that help determine the proper stocking policy for each item? Some of the factors include:

- What is the part?
- How critical is it to plant operation?
- What is its usage pattern?
- Is the part high usage? Low usage? Seasonal usage?
- What does the part cost? Is the part high cost? Low cost?
- Is there a discount for volume order?
- What is the part's stock out impact?
- How expensive is the downtime?
- How long will the downtime last?
- What is the part's lead time?
- How long does delivery take after the part is ordered?

Types of Spare Parts

Another way of looking at spare parts is by their part type. In any MRO storeroom, spare parts can be categorized by at least eight different types. These are:

- Bin stock—free issue
- Bin stock—controlled issue
- Critical or insurance spares
- Rebuildable spares
- Consumables
- Tools and equipment
- Surplus parts
- Scrap or useless parts

Bin Stock—Free Issue

Bin stock, free-issue items are typically parts like fasteners, including bolts, nuts, and washers. Pipe fittings may also fall into this category. These items are usually stored in bins located directly outside the storeroom, where employees can take what they need to perform their jobs without requisitions.

Bin Stock—Controlled Issue

Bin stock, controlled-issue items are items with a little more shelf value than free-issue items. These are items that generally need some control. They are usually behind the storeroom counter and issued to a requisition.

Critical or Insurance Spares

Critical or insurance spares are parts such as motors, pumps, gear cases, and other large spares. These are usually stored in locations where they can be carefully tracked and cared for so they are not damaged in storage.

Consumables

Consumables are maintenance-related items that are typically used in the performance of maintenance work. Items used during the performance of repair activities may include rags, tape, and speedy dry, etc.

Tools and Equipment

Tools and equipment are typically larger repair tools that are not assigned to a specific individual. These are issued to an employee or work order, then returned when the job is finished.

Surplus Parts

Surplus parts are found in the storeroom and are typically left over from projects or larger maintenance tasks. These surplus items should always be entered into the inventory tracking system before they are stored in the storeroom.

Scrap or Useless Parts

In most MRO storerooms, there are some scrap or useless parts. These items should be targeted for removal from the inventory system.

ABC Analysis

With this many types of items in the storeroom, how can the proper stocking rules be applied to each item? One tool that is useful in any MRO storeroom is ABC Analysis. This tool focuses on a significant few rather than the overwhelming many.

"A" Items

"A" items are typically of high dollar value, but low usage. These items may make up 80% of the total inventory costs. However, these items will also make up less than 20% of all the spare parts carried. "A" items should have regular reviews of reorder points and reorder quantities, as well as usage trends. All details concerning these items should receive close follow-up. It is important for them to have complete and accurate MRO inventory and purchasing records.

"B" Items

"B" items are of moderate dollar value and have moderate usage. These items may make up a total of 15% of the total inventory costs. However, they will typically only be 30% of the total inventory items. "B" items should receive regular cycle counts and tracking; attention should be paid to keeping good transaction records.

"C" Items

"C" items have a low dollar value. These items will make up about 5% of the total inventory value. However, they will make up approximately 50% of the total number of inventory items. "C" items will be the larger part of the inventory. Because they have less value, they will require minimal recordkeeping.

Being able to classify the spare parts in the storeroom into one of the ABC classifications allows the storeroom resources to focus on controlling and managing the most important items.

Order Quantity Rules

When considering the order quantity rules for any of the ABC items, there are at least three sets of rules that can be used. "A" items will typically be controlled by maximum and minimum quantity specifications. "B" items will typically use some type of fixed re-order interval system. They may be reordered weekly, monthly, or quarterly. "C" items typically use a two-bin system. Items are issued out of one bin until it is empty. Then they began issuing from the second bin; meanwhile the first bin is restocked. Utilizing these simple controls reduces the workload in any MRO storeroom.

Factors that Drive MRO Inventory Management

A question commonly asked is, "Should material cost be the primary factor driving MRO inventory management?" The answer is no. Various factors require adjustments due to any MRO inventory and purchasing disciplines.

Seasonal Demand

Among the most common factors are materials with seasonal demand. These items may be required for seasonal shutdowns, turnarounds, or outages. They typically show little or no usage during the majority of the year, but may have very high usage during the shutdown, turnaround, or outage period. In this case, typical inventory controls would need to be adjusted.

No Stock-Outs Allowed

In addition, there are materials for which stock-outs must never occur. These items are spare parts for critical equipment and processes in the plant. If there were a stock-out, the cost would have a dramatic impact on the company's profitability. Therefore, these items are probably kept at a higher stock level than routine spare parts.

Fabrication

Some spare parts must be fabricated. These spares will typically be capped at a higher stock level because the time to fabricate the spare part can be considerable. Additional stock is capped to reduce the amount of downtime that would be incurred if the part failed and no spares were kept.

Foreign Suppliers

Another consideration involves spare parts that come from foreign suppliers. In many cases, foreign suppliers only make spare parts during certain times of the year. If a stock-out were to occur, and the supplier was months away from making spare parts, the resulting downtime could be disastrous. Again, given this situation, it would be advisable to stock a larger number of spare parts. This factor should be a major consideration for any companies considering buying foreign equipment.

Quantity Minimums

A final thought about the impact of stocking policies when suppliers have dictated order quantity minimums. Although the plant may want to order only 10 of an item, the supplier may have a minimum order quantity of 50. The company will have to order 50, and this amount will show as excess inventory. Unfortunately, unless the company can negotiate with the supplier, it may always show stock overages with these types of spare parts.

Work Flows in the MRO Inventory Department

With these items in mind, the next step is to understand how work flows in the MRO inventory department.

Requesting Spare Parts

Figure 2-1 shows the flow process for requesting spare parts. It begins when work is in the planning process and a parts requirement is

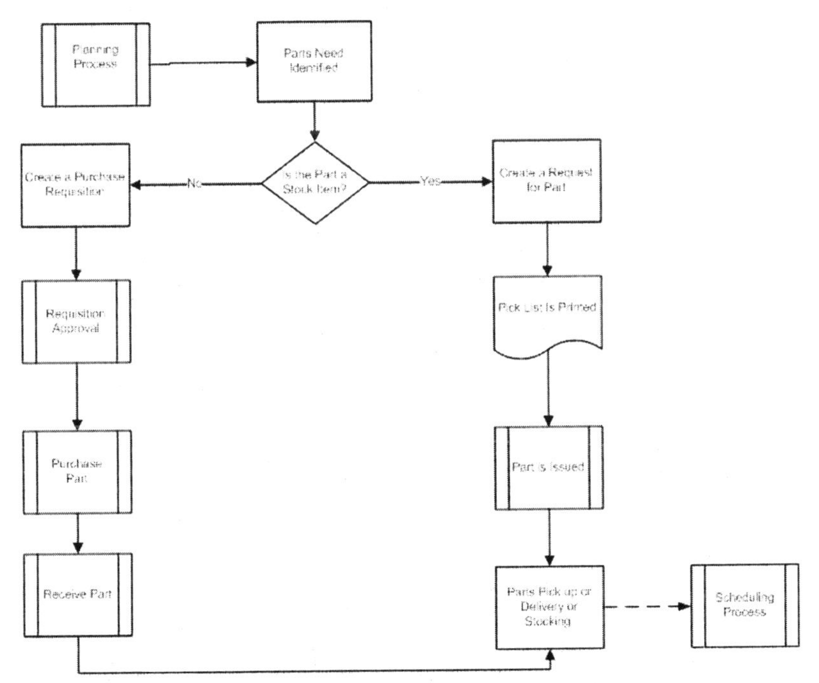

Figure 2-1 Requesting Spare Parts Process

identified. The planner will then notify the storeroom, usually electronically, that a part is needed. Once the storeroom clerk sees the request for the spare part, the question is asked, "Is the part a stock item?"

If the part is not a stock item, then the storeroom clerk will create a purchase requisition. The storeroom clerk will have the requisition approved, and an order placed for the part. The order is then sent to the vendor, who will process the order and ship the part. Once the part is received, the storeroom clerk will finish processing the order.

If the part is a stock item, the storeroom clerk will create a requisition for the part. This step is usually performed electronically. The part will be issued against the work order number that the planner had specified. After being notified of the weekly schedule, the storeroom clerk will print a pick list of all of the material items required for the schedule. Working from the pick list, the storeroom clerk will gather all of the spare parts in a staging location.

Inventory Disciplines

FIFO

- First in first out
- Used to ensure regular use of inventoried materials
- Help with reducing obsolescence, excessive shelf life and deterioration

LIFO

- Last in first out
- Primarily used for improved financial reporting during periods of rapid inflation

FISH

- First in still here
- A common inventory discipline used where no disciplines are practiced

At this time, the storeroom clerk will process the part as being issued against the work order. When it is time for the maintenance technicians to perform the work activity, they simply stop by the storeroom window, provide the storeroom clerk with the work order number, and are issued the spare parts they need.

Issuing a Spare Part

As shown in Figure 2-2, there are three possible sources for the request of a spare part. The first, as was previously mentioned, is a demand from planned work. A second source of request can be an unplanned job. The third source can be a contractor or a project activity. Regardless of the way the request is created, it is processed very similarly.

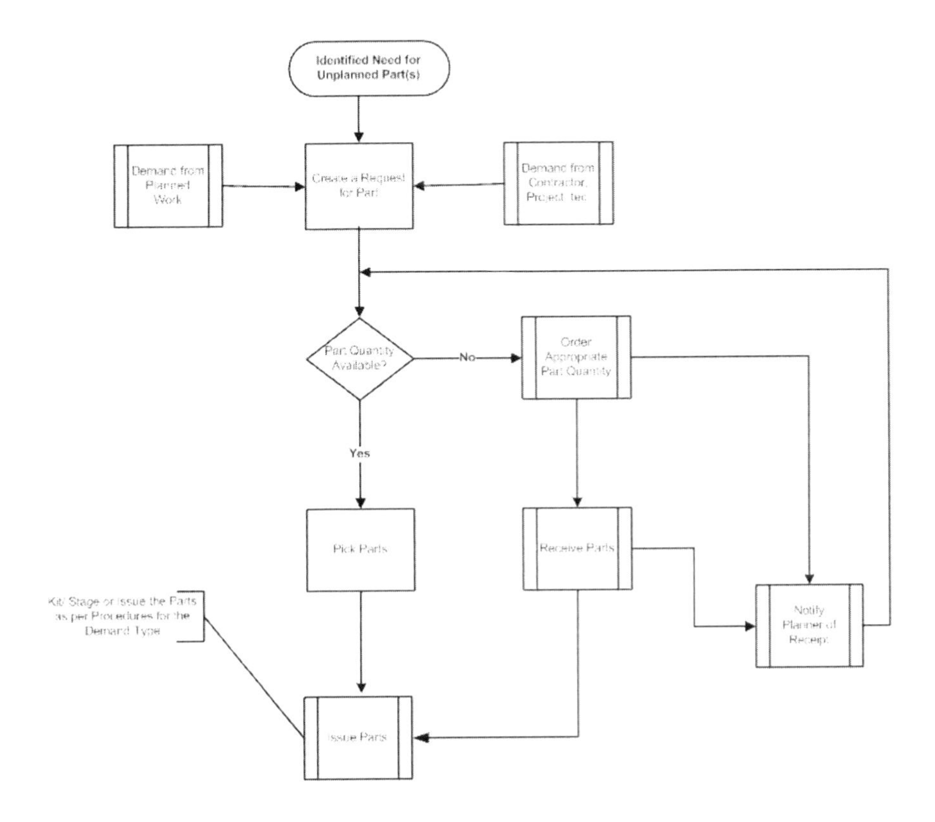

Figure 2-2 Issue of Stock Part Process

When the request is received, usually electronically, the storeroom clerk will check to ensure that the item is in stock. Once the item is confirmed as a stock item, the storeroom clerk will check that the proper quantity of spare parts are on hand and available to issue.

If the part quantity is available, the storeroom clerk will pick the parts, stage the parts, and prepare them to be issued. If the spare part requested is for an unplanned job, the issue may take place immediately.

If the part demand is from a contractor or project, the part may not be issued immediately, but rather when the work is scheduled and performed. Similarly, if the demand came from planned work, the part will be staged and set for delivery.

If the part quantity is not on hand, then an order must be placed. The individual requesting the spare part will be notified either verbally or electronically that the work cannot be performed due to a parts shortage. A probable delivery date for the spare parts will be provided so that the work can be rescheduled. Once the ordered parts have been received, the planner or the individual requesting the spare part will be notified of the receipt; the work can then be rescheduled. This part of the process is shown in Figure 2-2 as a feedback loop from notifying the planner and insuring that the parts are available.

Free Issue Items

In some cases, spare parts are not kept behind the storeroom counter. Instead, they are typically stored in front of the storeroom, where employees can take the number of items that they need to perform their work activities. These items are typically parts such as nuts, bolts, washers, and other low-cost, high-volume parts. They are known as free issue items (see Figure 2-3).

Once a part is defined as a free issue item, a vendor is established for the spare part. The storeroom personnel will work with the vendor to establish certain stocking policies and levels for these free issue items. As employees withdraw free issue materials from the available stock, it eventually drops below a minimum level. The vendor will visit the site on a preset schedule, come to the stores location, and count the available spare parts. If the on-hand quantity is below the agreed to amount, the vendor representative will restock the item.

After restocking the free issue items, the vendor will notify the storeroom personnel of the replenishment quantity and cost. The storeroom personnel will then receive the free issue parts and add the cost to the storeroom valuation. This cost is typically charged to a standing purchase order or requisition for tracking the cost.

The storeroom personnel will randomly check the vendor's restocking quantities and usage patterns to ensure that the vendor's stock counts are accurate.

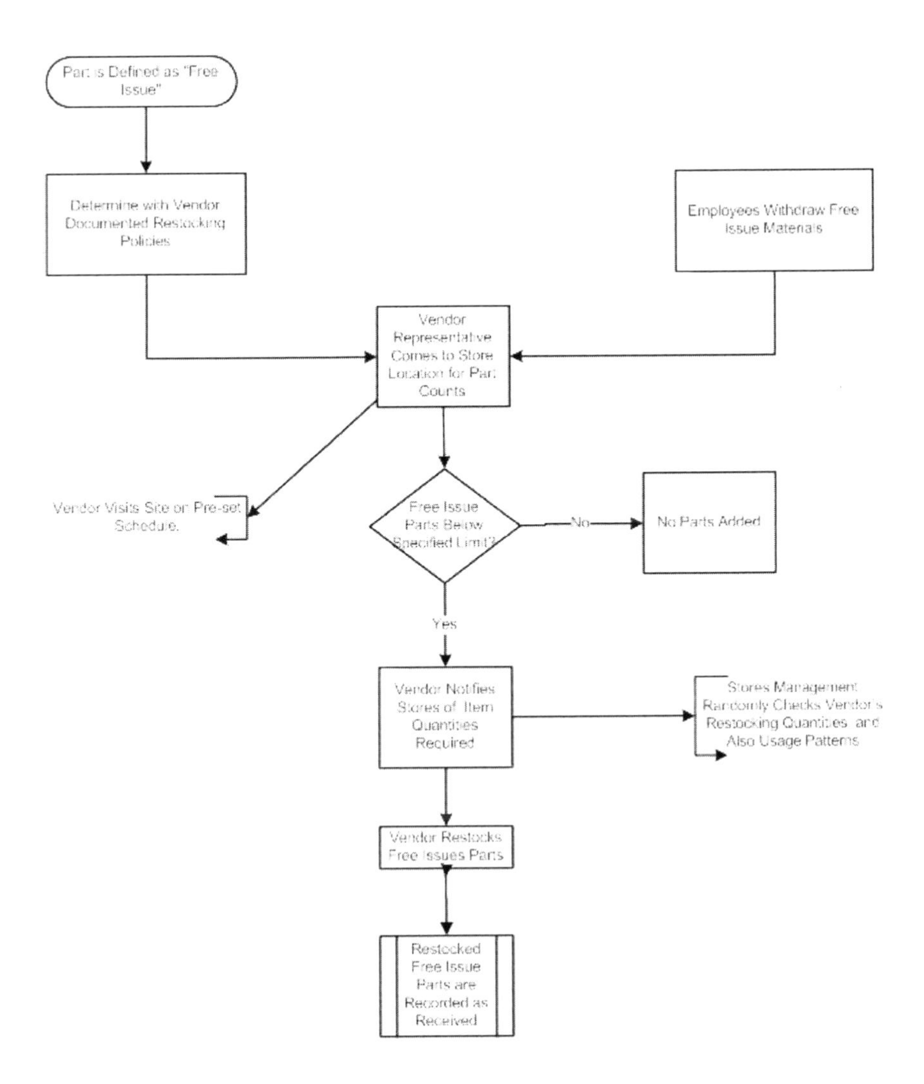

Figure 2-3 Replenishment of Free Issue Part Process

If any personnel needs free issue items of a substantial quantity, the storeroom personnel should be notified so that a larger charge can be debited to the work order on which they are working. It should be up to the company to set the policy of what the dollar value should be for this type of transaction. Typically, free issue items are used where the cost of processing a requisition is greater than the cost of the item.

Direct Purchase Items

Direct purchase items are spare parts that are used infrequently. This means it is not justifiable from a cost perspective to carry them in inventory. Because they will be ordered on an "As Needed" basis, a separate flow for procuring these items is needed.

The flow for direct purchase items is detailed in Figure 2-4. It begins with the need for the non-stock item or service identified. Because an individual will have identified the need for the item, it will be up to this individual to complete the requisition, or at least the request to begin the direct purchasing process.

The requisition or purchase order will then progress to the individual making the decision. Based on the price of the item, this may be a one step or multiple step process. If the purchase is approved, it proceeds to the next step. If the purchase is not approved, it is sent into a re-evaluation loop, where more information may need to be gathered to justify the purchase. It is possible at this point to reject the direct purchase.

If the direct purchase of the spare part or the service is approved, it is next moved to the point where a vendor (for a part) or a contractor (for a service) is contacted. If it is for a contracted service, the competing vendors are evaluated and the best candidate is selected. Once any vendors are selected, they are notified and integrated into the scheduling process.

If the direct purchase is for a spare part, the competing vendors are evaluated, based on price and delivery service. The part is then ordered from the selected vendor. Once the part is received at the stores receiving, the requestor is notified. The planner is then notified and the part is bagged or tagged and made available for the work order that created its demand. The work is then executed and completed.

Whether an item is stocked, free issue, or direct purchase, it needs to be controlled. The suggestions in this chapter should help an organization to develop the proper control procedures to minimize inventory costs and increase the overall service level.

3

Tool Management

Stocking Tools

Most MRO storerooms also contain maintenance and repair tools. Typically, these tools are not assigned to any one technician, but are kept in a central location so that all technicians have access to them. This category also includes tools that are not used frequently. Therefore, there is little likelihood that the tools will be demanded at the same time. These tools are typically high dollar, expensive to repair, or replacements for lost or damaged tools.

Tracking Tools

Figure 3-1 shows the typical process for controlling tools. A demand is usually issued from planning for a particular job, which is assigned a work order number. The plan typically will identify the tool that is required for the job. A request for the tool, usually generated electronically, is sent to the storeroom. Depending on the scheduling window, the tool may be set aside for the particular work. If the tool is not set aside, the storeroom is usually notified when the weekly schedule (pick list) is run, and the tool is staged with the spare parts for the same job.

When the tool is issued, it is typically issued to a technician, whose ID is placed on the work order. If the work is being performed by a contractor, then some identification from the contractor is typically left at the storeroom when the tool is issued. This is to ensure that the tool is properly tracked, whether it is issued to an internal technician or a contractor.

When the work is finished, the tool is returned to the storeroom. The work order that the technician or contractor was working on is clearly identified. The technician or contractor then turns the tool into the storeroom attendant. The storeroom attendant checks the tool to see if it

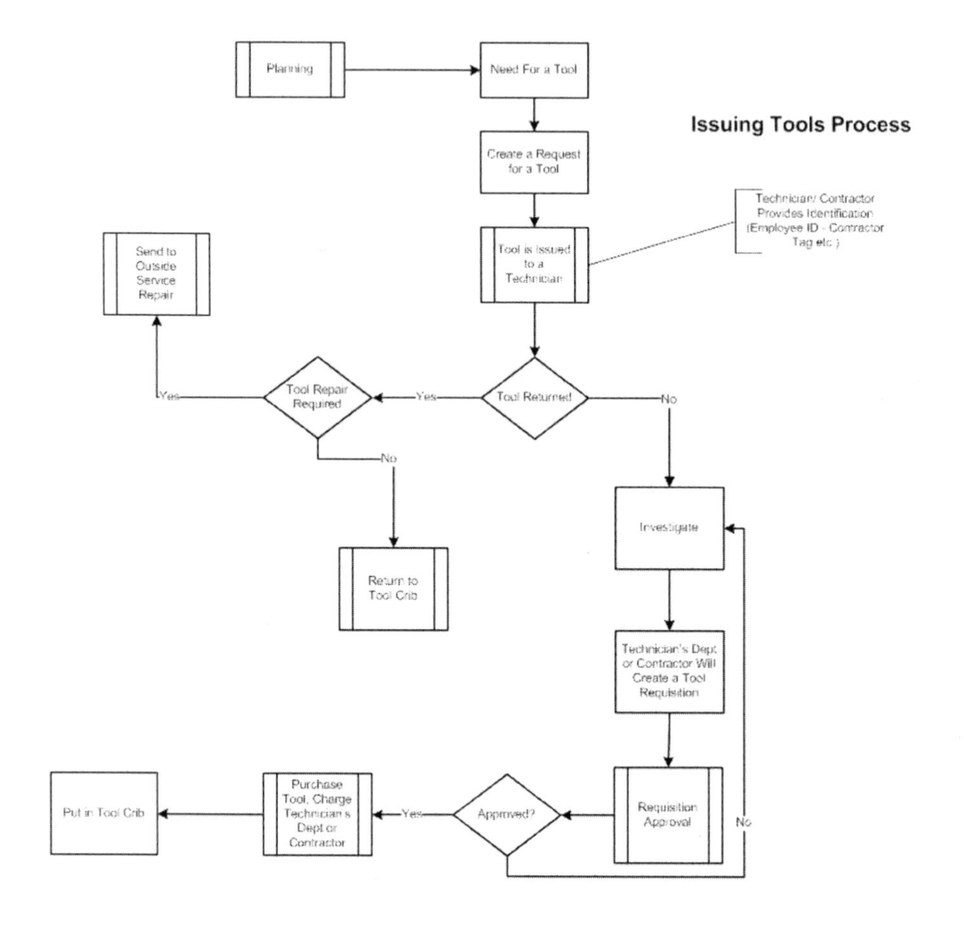

Figure 3-1 Issuing Tools Process

has been damaged. If the tool is not damaged, then the storeroom attendant returns it to the tool crib. If the tool has been damaged, then a decision is made whether the tool will be repaired internally or be sent to an outside service for repair. Once the tool is repaired, whether internally or by the outside service, it is returned to the tool crib.

Replacing Tools

If the tool is not returned, then the matter is investigated. The technician's department or the contractor will issue a requisition to purchase a

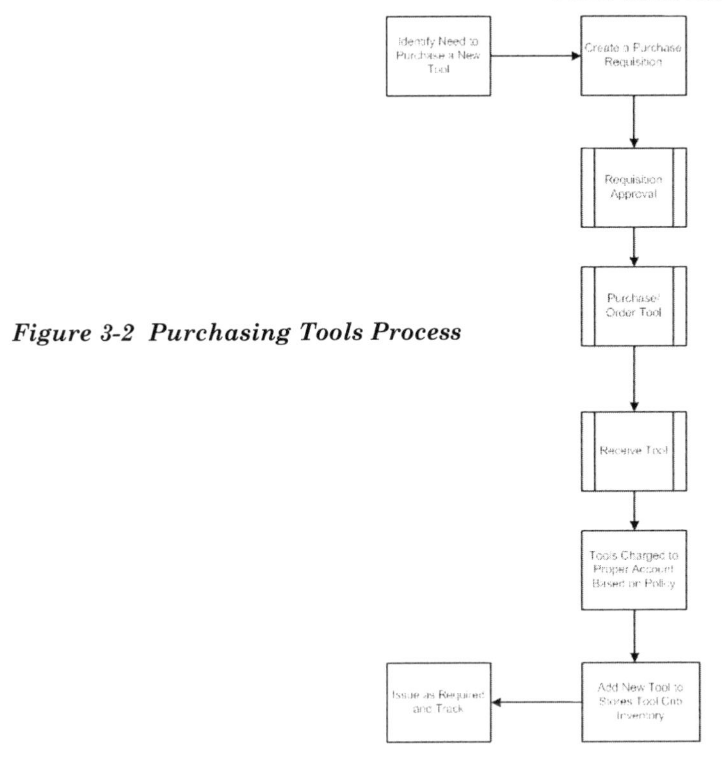

Figure 3-2 Purchasing Tools Process

new tool. Once the requisition has been approved, then the tool is purchased. The price of the tool is charged to the technician's department or to the contractor. Once the new tool has arrived, it is placed in the tool crib by the storeroom attendant.

If the requisition to buy a new tool is not approved, then the matter is reinvestigated, until a resolution is reached. The final goal is to have a new tool placed in the storeroom to replace the one that was lost. This ensures that the tool will always be in the storeroom when it is needed for a job.

Purchasing Tools

A more detailed description of purchasing tools is illustrated in Figure 3-2. A need to purchase a new tool is identified to start the process. This need can arise from the fact that a tool has been lost or, perhaps, a new way of performing work has been identified requiring the purchase of a special tool.

Once the need to purchase the tool has been identified, a purchase requisition is created. The requisition is then sent for approval. The approval may come from the storeroom, from the maintenance department, or even from the accounting department. It is typically a matter of company policy that dictates which department approves purchasing a new tool. Once the tool is approved, the storeroom attendant places an order for the tool.

When the tool is received, it is checked to ensure that it is in good condition. Once deliveries have been accepted, the cost of the tool is charged to the proper account, based again on company policy. The storeroom attendant then adds the tool to the stores inventory crib, and enters the information into the computerized inventory system. The tool is then ready to be issued for work.

Specialized tools can make up a considerable investment for many companies. Utilizing a tool room control system as part of the MRO inventory and purchasing process can help to minimize this investment.

4

The Reordering Process

As the storeroom stock is utilized, it becomes necessary to restock items. There needs to be a process for reordering items, whether they are stock or non-stock items. This chapter will concentrate on stock items.

Ordering Stock Items

Figure 4-1 illustrates the process of ordering stock items. As spare parts are used from the storeroom, the stock is depleted. Based on certain criteria—for example, max-min levels or a type of reorder point—the stock will be reordered. The items to be reordered will typically be identified by a computerized inventory system. These systems can range from very simple to very complex systems. Still, even the most basic system will create a need-to-reorder report.

Once the storeroom attendant has run this report, it is necessary to identify whether the parts needing to be ordered are pre-approved for purchase or must be approved individually. If the part is pre-approved for purchase, the storeroom attendant will create a purchase requisition. Once this requisition is approved, it will be issued as a purchase order, which will then be sent to the appropriate vendor.

If the part is not pre-approved, a report should be printed and reviewed line item by line item. Depending on the authority level required, the storeroom attendant, a manager, or a designated approver will make the decision to reorder each part. These individuals will be provided with certain criteria on which to order each item. If the order falls within the criteria for the stock item, then the parts are purchased. The purchasing process then becomes similar to the pre-approved items process.

However, if the reorder criteria is not met, then the individual line item will be excluded from the order. A business case will need to be developed and presented to the next higher level of approval to be able to

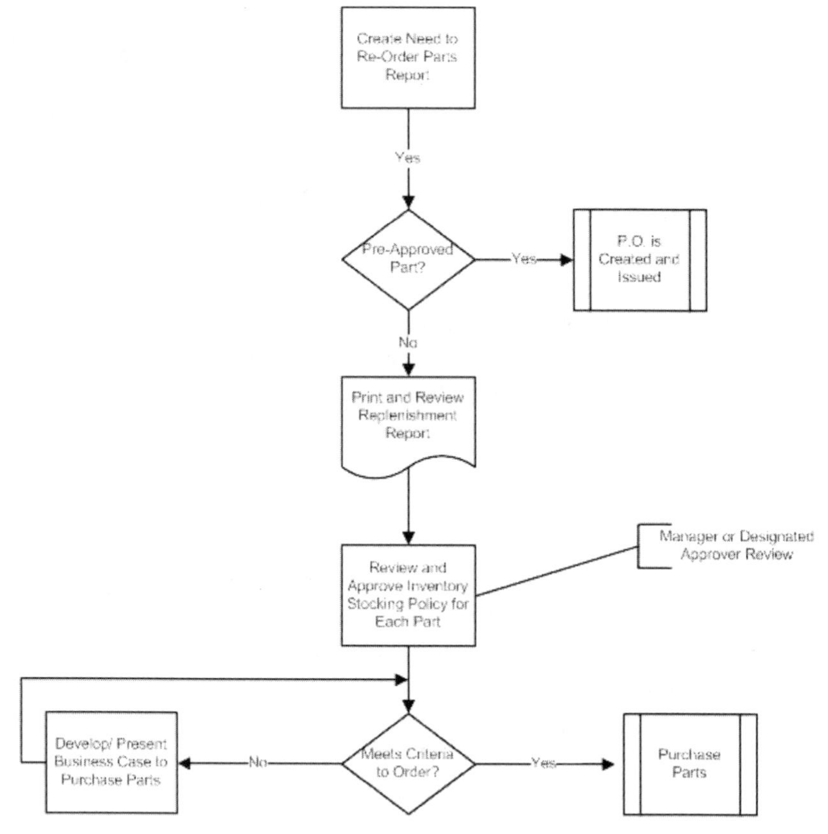

Figure 4-1 Inventory Re-order Process

purchase the parts. Once this approval level is satisfied, the order can loop back through the ordering process.

The Approval Process

The approval process for some companies can be quite complex. Figure 4-2 highlights a common approval process. This approval process is for parts and services. It begins by examining whether or not parts or services are required. If they are required, the request must specify if the work is a breakdown or emergency or if it is routine work. If the work identified is an emergency, approvals are usually authorized after the repair has started. In the case of a breakdown or emergency, the maintenance supervisor on duty is typically authorized to purchase the material

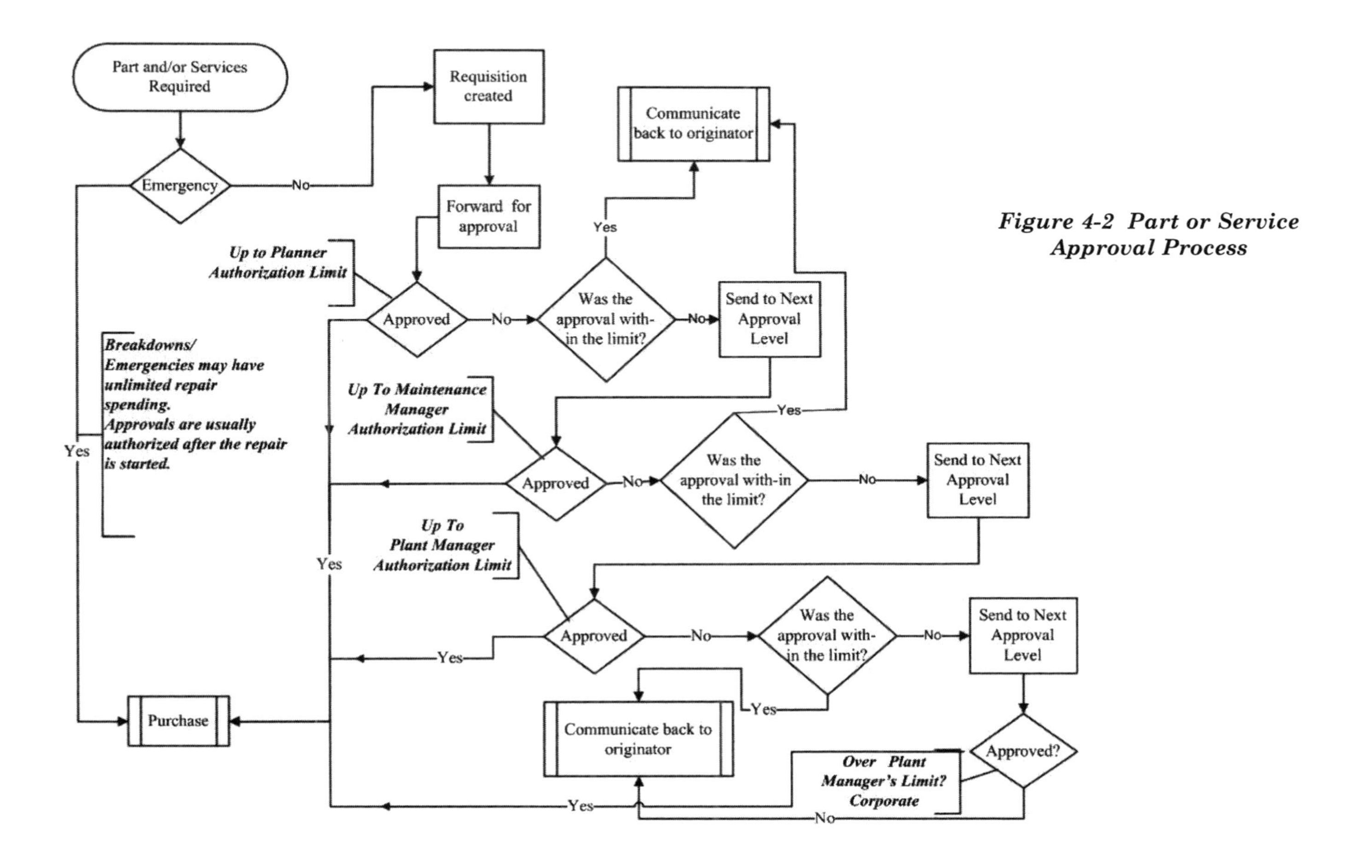

Figure 4-2 Part or Service Approval Process

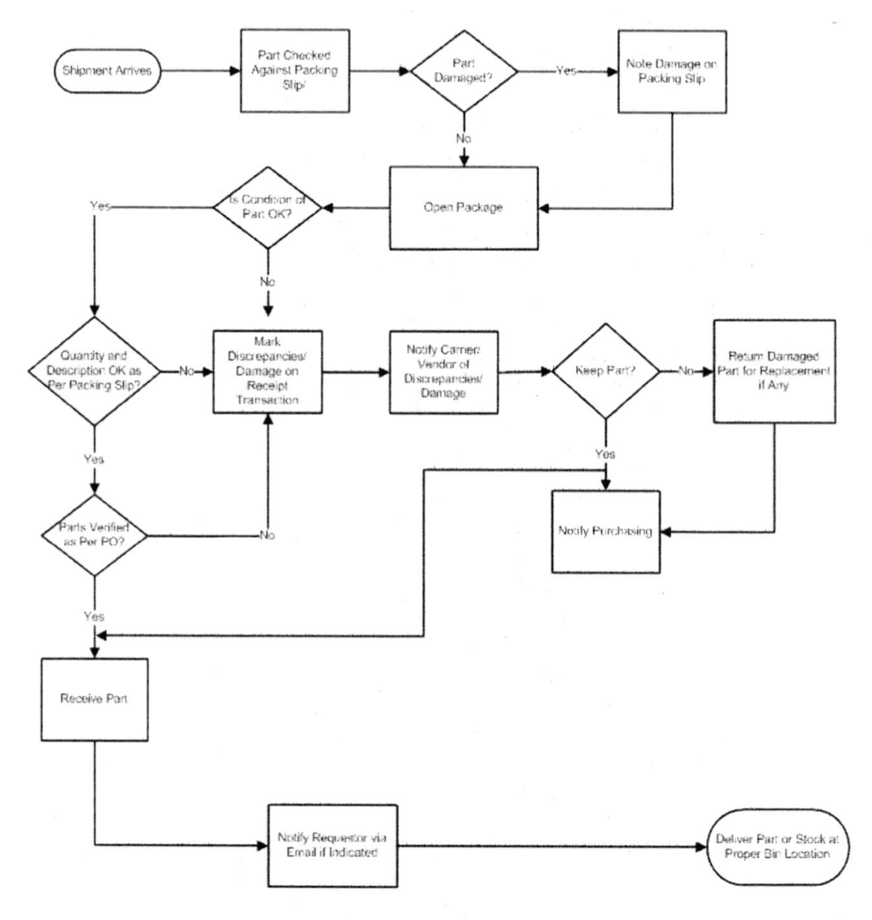

Figure 4-3 Spare Part Receiving Process

or service necessary to restore the equipment to an operating condition.

If the part or service required is for routine work, a requisition is created. Maintenance planners will usually be the individuals responsible for creating the requisition, as part of the planning process. Most computerized maintenance management systems or enterprise asset management systems will generate this requisition automatically as part of the planning process. Once the requisition is created, it is forwarded to the appropriate personnel for approval. In many cases, the planners themselves have the appropriate approval level for the spare part. This level is usually established as part of the company policy, with a limit at a certain dollar amount. If the required part's requisition is within the planners' approval level, then it is approved and the order is placed.

There may be reasons other than financial that cause a purchase not to be approved. It may be that the work is not going to be performed at the present time or the job is deemed superfluous. If this is the case, then there should be some communication back to the originators of the request to explain why the work is not being done. If they still feel the work is important, then they may escalate the request for the spare part or the work to be performed. If the purchase was not within the planner's approval limit, then it may be sent to the next approval level, which is typically the maintenance manager.

The maintenance manager will be authorized at a higher spending limit than the planner. The maintenance manager has a choice to approve the purchase for the part or service or to deny it. If the purchase is approved, then the requisition is forwarded on and the part or service is ordered. However, the maintenance manager may decide that the work is not to be performed and turn down the request. If this is the case, the decision should be communicated back to the originator who, again, then has the opportunity to resubmit the request.

In some cases, the maintenance manager's authorization limit is not sufficient to approve purchasing the part or service. In this case, the appropriation will be sent to the next approval level. In most organizations, this would be the plant manager's level. The plant manager may decide the work is necessary and authorizes the purchase of the part or service. On the other hand, the plant manager may also decide the work is not necessary, and deny the request. As before, if this is the case, there should be some communication back to the originator, who can modify or escalate the request as necessary.

In rare cases, the requisition for the parts or services may exceed the spending limit of the plant manager. In this case, the request would need to be escalated to a corporate representative. This representative would then have the choice to approve the requisition for the parts or service, or to deny the request. If the request is denied, communication back to the originator is essential. At this level, however, it is most likely that the plant manager believes the approval should be given; the corporate representative will likely take this into consideration. If the corporate representative approves the purchase of the parts or service, then the requisition is typically sent back to the plant to be properly executed.

The Receiving Process

Once an order has been placed, and the vendor has processed it, the part is shipped to receiving. A distinct process should be followed when a part is received. A common flow for spare part receiving is pictured in Figure 4-3. This flow begins with the shipment arriving at receiving and the receiving clerk checks the part against the packing slip.

The first thing the receiving clerk looks for is any damage to the part. Any damage should be noted on the packing slip. The receiving clerk then opens the package for further inspection. Once the package is opened, the part is again examined for damage. The receiving clerk should mark any damage on the receipt transaction. The receiving clerk then notifies the carrier and the vendor of any damage. The decision will then be made whether to keep the part order or to return the part to the vendor.

If the condition of the part is good, then the receiving clerk checks the quantity and the description of the part, comparing it to what was ordered on the packing slip and eventually the purchase order. Any discrepancies should be noted on the receipt transaction; again the carrier and vendor should be notified of the discrepancies. As before, a decision is made as to whether the part should be kept, or returned to the vendor.

If there is a discrepancy or the part is damaged, and the spare part is returned to the vendor, then purchasing should be notified immediately. The purchasing department will determine how to re-order the spare parts. It will also be in a better position to renegotiate the pricing of these spare parts due to the inconvenience of receiving a damaged or incorrect order.

If the quantity and description of the spare parts match the packing slip and the purchase order, the part can be received. This process will vary from company to company, but it should involve electronically noting the receipt, and closing out the purchase order. The spare parts can then be placed in the appropriate storage location. This location also should be noted in the electronic inventory system.

If the part was special ordered, rushed, or needed for a project, it may be necessary to arrange for a delivery to be made. This delivery can be to a job site, to the maintenance office, or to the project's storage and staging location. Any delivery needs should be specified on the requisition, and the purchase order.

Good process flows, coupled with good business processes, can help companies lower their restocking costs.

5

REBUILDABLE SPARES

Many organizations have spare parts that are so large and complex that they can be rebuilt. In some cases, these may be components such as pumps, motors, gear cases, or other major spares. Because these components are to be repaired or rebuilt, they follow a specialized process. A basic rebuild process is highlighted in Figure 5-1.

This flowchart starts at a point after the component has failed and been replaced. The first question planners should ask is, "Is the part repairable?" If the answer is no, and the planners still need to have the part rebuilt, then they need to change the status of the part to show it is a rebuildable component.

Once the part is identified in the electronic system as a rebuildable spare, all approvals or purchase orders necessary to have the item rebuilt should be obtained. If the component is to be repaired internally, it is likely that the appropriate approval will be all that is needed. If the component is to be repaired externally, by a contractor, then it is likely that a purchase order will be required for the repair.

Rebuilding Internally

If the component is to be repaired internally, the planners should first check to ensure that all spare parts are available. If the spare parts to rebuild the component are not available, the planners should order the spare parts.

Once the spare parts are available, the planners should issue a work order to rebuild the component. The work order should be used to charge all of the labor and materials that are required to rebuild the component. This will ensure that the true costs to rebuild the component will be captured.

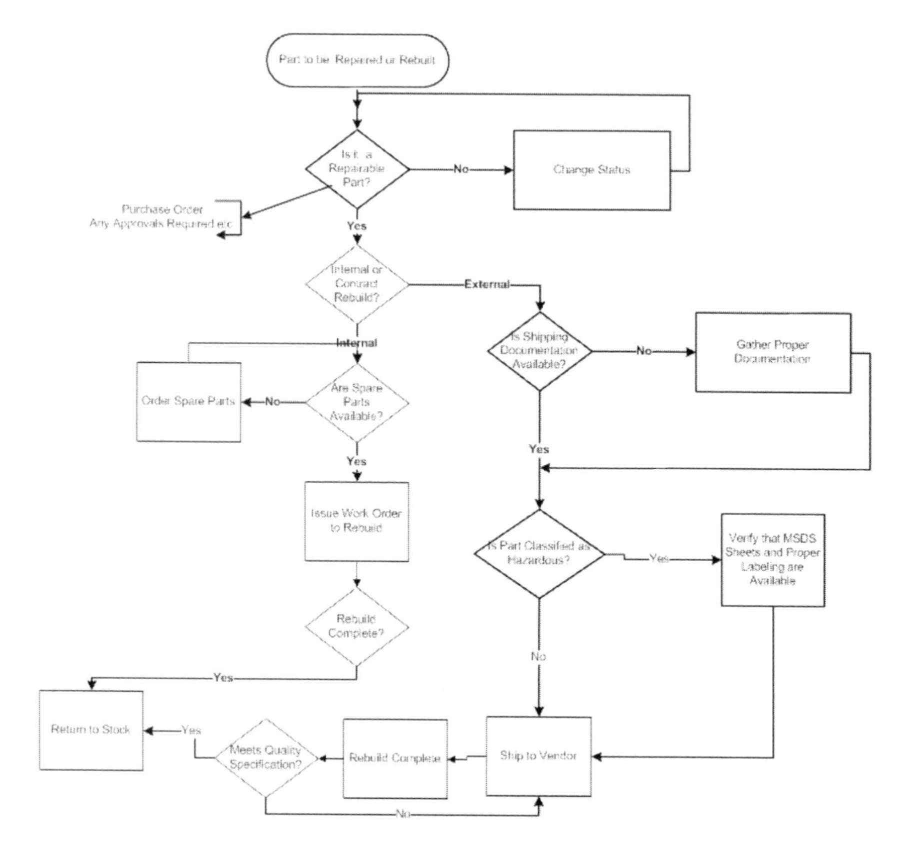

Figure 5-1 Repair/Rebuildable Part Process

Once the item is completely rebuilt, and the quality checked, it should be returned to stock.

Rebuilding Externally

If the component is to be rebuilt externally, then it needs to be prepared to ship to the vendor. This entails gathering the appropriate shipping documentation and insuring that it's available when the item is picked up. If the component contains hazardous material, then the appropriate MSDS sheets and proper labeling should be available prior to shipping. Once the appropriate paperwork is prepared, the component is shipped to the vendor.

In turn, once the vendor has completed work on the component, they notify the appropriate representative at the plant. Depending on the type of rebuild, someone from the plant may need to visit the vendor to do a site inspection, ensuring the quality of the rebuild. In other cases, it is appropriate for the vendor to ship the rebuild to the plant, with the quality acceptance inspection being performed at receiving. If the component meets the quality specifications, it is returned to stock. If it fails to meet the quality specification, then it is returned to the vendor for proper rebuild.

Issues with Rebuilds

One of the major problems with rebuilds is pictured in Figure 5-2. The problem deals with the costing of rebuilds. For example, in Figure 5-2, a motor has been installed and has a value of $5,000. If the motor fails, it is sent to the rebuild shop, and rebuilt. The total cost to rebuild the motor is $1500. When the motor is put back in to storage, what is the new value of the motor? In some companies, you will find that the motor is worth:

- $5,000
- $1500
- $3500
- $6,500
- $0.00

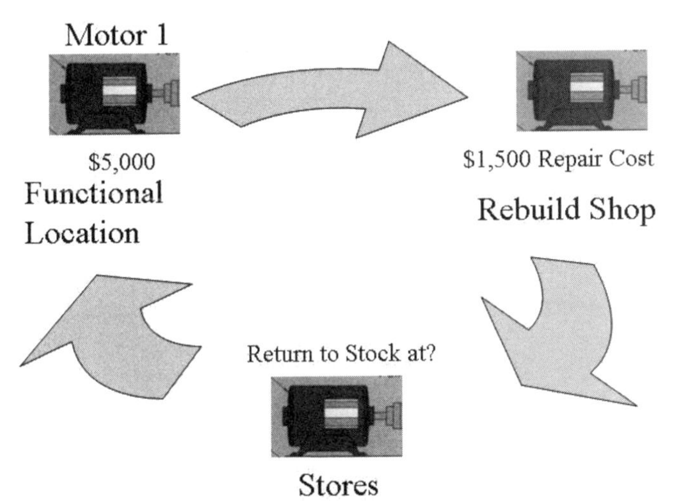

Figure 5-2 Rebuild Costing?

An individual may wonder why the way the costing of a rebuild is charged is considered important. Consider the diagram presented in this book's Introduction, in Figure I-4. In this diagram, the inventory and purchasing system are linked to the work order, and eventually to an equipment history. The material costs are passed from inventory to the work order to the equipment history. If a large variance occurs in the costing, then the cost posted to the piece of equipment is inaccurate or inconsistent. If a detailed life cycle cost analysis is performed on equipment and cost variances are allowed to affect the material charges through the rebuilds, how accurate can the analysis be?

Consider a hypothetical situation where we have a new motor worth $5,000 in stores and an old motor still functional, worth zero dollars. If you were responsible for their maintenance budget, which motor would you use to replace a failed motor? It can be quickly seen, that the valuation of the spares can have a dramatic impact on repair and replacement policies.

Rebuildable spares are a major investment for most capital-intensive companies. Using good process flows and a disciplined business process can help a company minimize these costs.

Returning Items to Stock

When parts are issued to a work order, and the technicians remove the spare parts from the storeroom and take them to the job site, there are occasions when some spare parts are not used. How are the technicians to return the items to the storeroom? There needs to be a process in place to allow excess items to be returned to the storeroom. Furthermore, these items must be properly credited to the work order and, ultimately, the equipment to which they were charged when they were issued.

The Return-to-Stock Process

Figure 6-1 highlights a return-to-stock process. This process begins when an unused part is identified to be returned to stock. The item is brought back to the storeroom attendant, who will need to know the work order the item was issued to. This step allows the attendant to adjust the work order so that the cost is properly reflected.

Controlling Stock Overages

The overage issued to the work order can create a cost problem in the storeroom. For example, if the item is returned, and a part is already on order, then the on-hand quantity will likely be over the maximum. If the item is returned and a part is already on order, the status of the outstanding purchase order should be examined. If the part is on order, but has not been shipped by the vendor, the order should be canceled unless unreasonable charges will be incurred for canceling the order. If the item has already been shipped by the vendor, it should be returned to the vendor, unless unreasonable charges would be incurred. If the vendor is going to institute unreasonable charges, an appropriate representative from purchasing should contact the vendor and review the relationship.

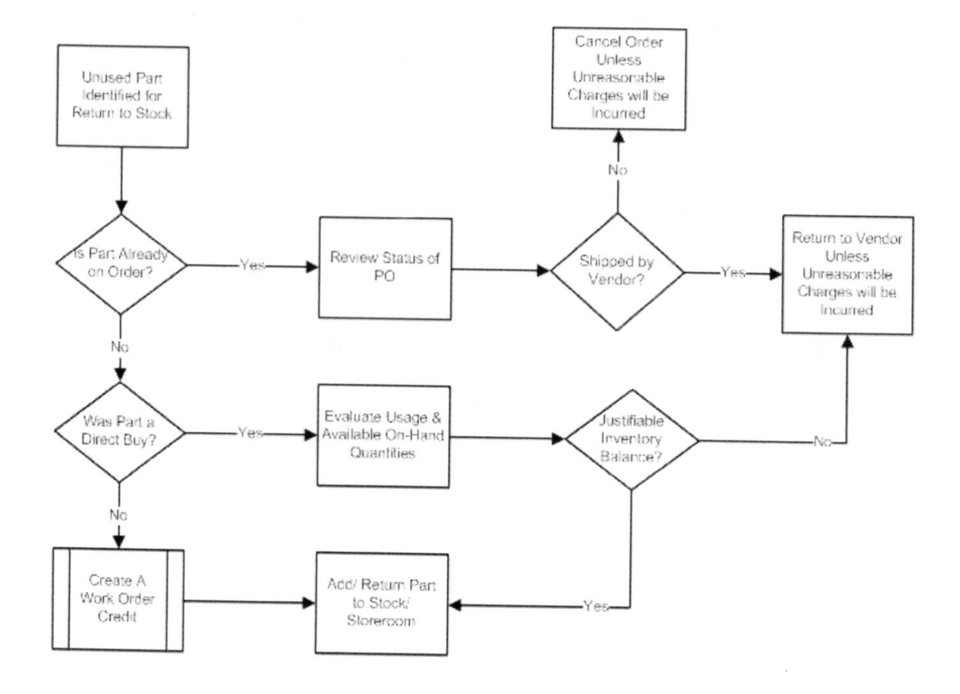

Figure 6-1 Returning Parts to Stock/Storeroom Process

If an item is being returned to stock, and the part is not already on order, the storeroom attendant should consider whether or not the part was a directed buy. A direct buy will typically occur for project or contract work. Rarely will a direct buy occur for a normal planned and scheduled work order. If the part is a directed buy, the storeroom attendant should evaluate current usage patterns, and also available on-hand quantities. If the part is a high usage item or if the on-hand quantities were low, it may be justifiable to add the item to inventory.

If the item being returned to stock was not a directed buy, the storeroom attendant should create a work order credit, then add or return the part to the storeroom. If, however, adding the part to the storeroom would exceed a justifiable inventory balance, the vendors should be contacted to see if they will accept the returned part. If the vendors will not accept the part without an excessive return charge, it may still be best to add the item to the stock room. If the vendors do impose an excessive return charge, someone from the purchasing department should contact them to review their relationship to the company. The company may do well to begin searching for other vendors for the spare part.

Returning spare parts to stock can help a company lower their annual material usage. However, it is important to understand the process clearly in order to avoid carrying too many spare parts and inflating the value of the inventory.

7

Adding and Deleting MRO Parts

Over time as a company adds and removes equipment from plant, it will be necessary to add spare parts to the storeroom and also to delete spare parts from the storeroom. It is necessary to have a defined process to add and delete spare parts.

Adding Spare Parts

When new equipment is purchased, the vendor will recommend spare parts that should be stocked. The list of spare parts should be provided to the storeroom attendant, who can check the list to see if any of the spare parts are already stocked in the storeroom. This check will prevent any duplicates from being carried in the storeroom. The consideration that the storeroom attendant should give to adding spare parts is highlighted in Figure 7-1.

As mentioned previously, the first consideration for storeroom attendants is to ask whether the part is already in stock. If the part is in stock, they should reevaluate the request, and also inform the individual requesting the item to be stocked that the item is already carried in stock. Rather than adding a new line item to the inventory, the attendants may simply need to adjust the maximum and minimum quantities carried in stock.

If the part is not already carried in stock, the storeroom attendants evaluate if the part is available from already approved vendors. Approved vendors are those with whom the company already has spare part arrangements. If no approved vendors carry the spare part, the storeroom attendants locate and approve a new vendor. Although the storeroom attendants probably can not complete this task on their own, they can at least make recommendations of the vendors who do carry the new spare parts.

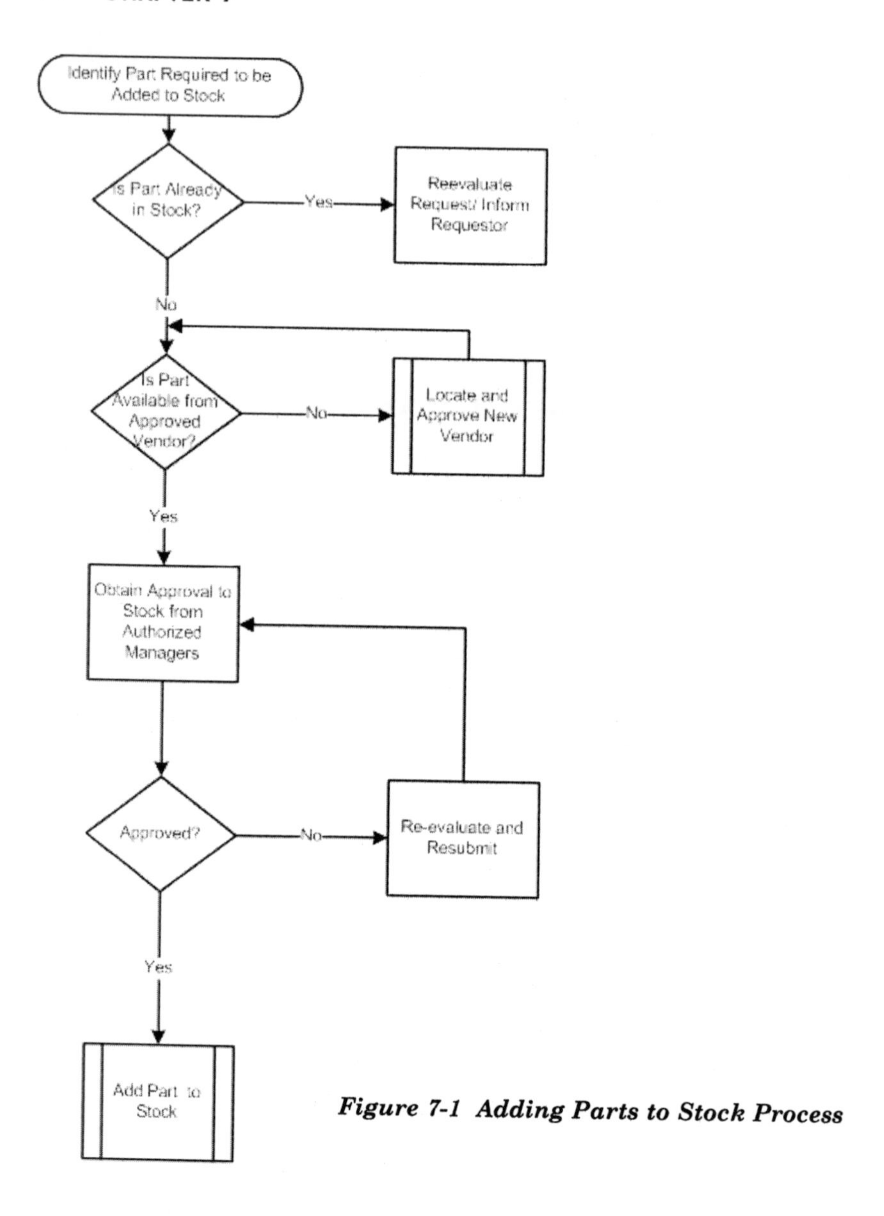

Figure 7-1 Adding Parts to Stock Process

If the part is already available from an approved vendor, then the storeroom attendants should submit a request to "add the item to stock" to the appropriate manager for approval. This approval process is important to prevent the addition of unnecessary items to the storeroom. Once the managers have reviewed the request, they may approve it. If the request is approved, then the appropriate part can be added to the stockroom. This task entails gathering all the appropriate information that was specified back in Chapter 1, and insuring that the proper stocking levels are set.

If the part is not approved by the manager, then it is up to the storeroom attendants to contact the requester; both the attendants and the requester should review the request and reevaluate it. They may decide not to stock the spare part, but rather set up a consignment arrangement with a vendor. If they still decide that the part needs to be in stock, then they should develop the appropriate business justification for stocking the spare part and resubmit it to the manager.

Deleting Stock Items

When equipment or a plant is decommissioned, it is financially beneficial to remove unnecessary spare parts attached to that equipment from the storeroom. Figure 7-2 highlights a deleting stock item process. When considering Figure 7-2, the spare parts data, particularly the historical usage and the "where-used" information, is critical.

This is highlighted in the second block, "Review Part Information". For example, some of the data required to be reviewed include the last time the part was used, the last date it was used, where it is or was used, a historical usage trend (usually for a year), and calculations for mean time between failure and mean time to repair. This data should not be reviewed by the storeroom attendants alone. As users of the stocked part, the maintenance department should be consulted before any deletion is allowed. They may have an understanding of information regarding the part that is not available to the inventory and purchasing departments.

Reviewing the data is for the sole purpose of determining if this part will be used in the plant any longer. If there is a potential for the part to continue to be used in the plant, it should be kept. The on-hand quantity and the reorder quantity should then be reviewed. Rather than eliminating the part, reducing the number of parts that are kept could be an alternative decision.

If the part is no longer going to be required in the plant, and there are items in stock, the vendor should be contacted to see if any of the items could be returned. In some cases, particularly if the company's relationship with the vendor is good, the vendor will be willing to return the item for credit. If this is the case, the item should be returned to the vendor and the vendor given a preferential status.

If the vendor is unwilling to receive the item or credit, then it would be best to remove the item from stock and either dispose of it as scrap or turn it over to an auction house to be sold. Before either decision is made, the parts should be discussed with the maintenance department who may have an alternative usage better than selling the parts for scrap.

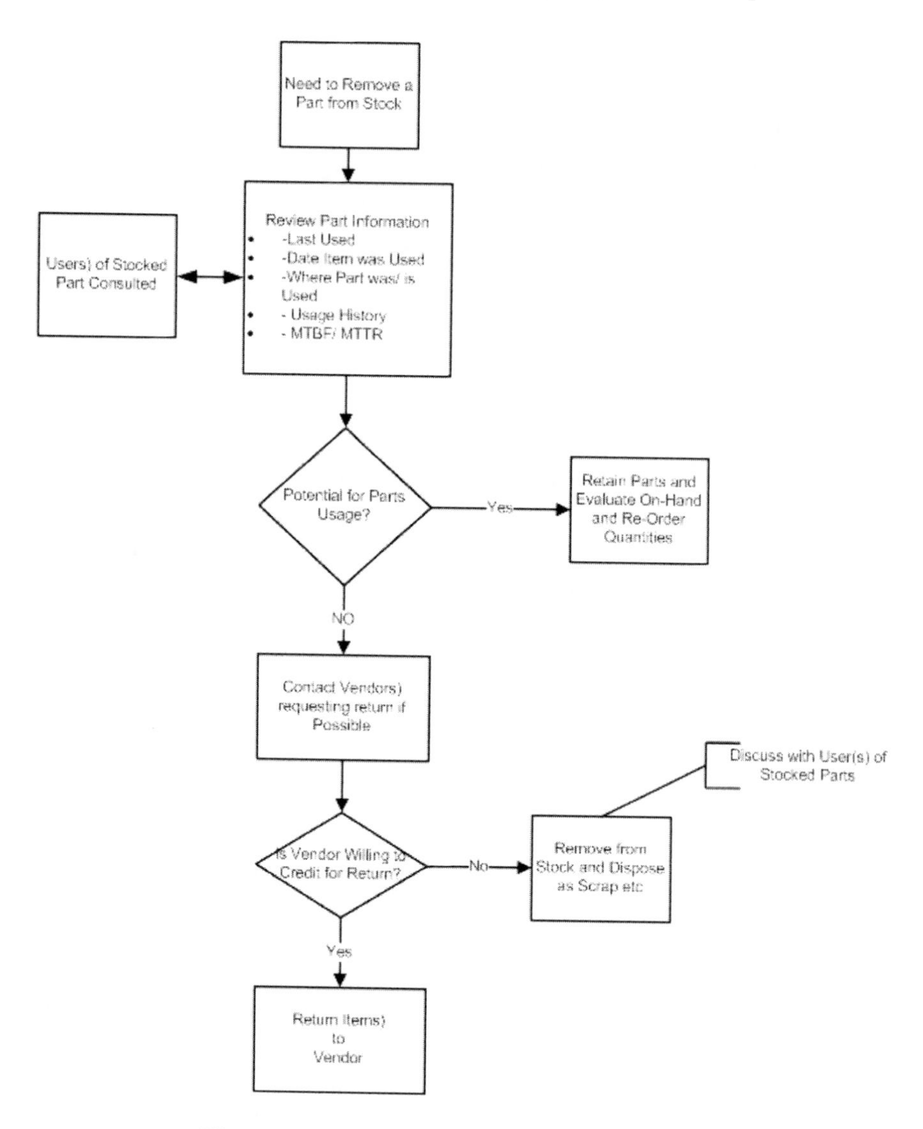

Figure 7-2 Deleting a Stock Item Process

Adding and deleting spare parts from inventory may seem like a simple function. However, unless good process flows are coupled with a disciplined MRO organization, the inventory levels can quickly become skewed, having a negative impact on the company's business and profitability.

8

Making Financial Adjustments to MRO Inventories

As inventories are used, variances will occur from time to time. This is particularly true when storerooms are uncontrolled and unmanned during off shifts. Spare parts will get used, and the usage is not always properly documented. This leads to spare parts no longer appearing in inventory, but without a clear place to debit the cost. When this occurs, an inventory valuation adjustment is required. Figure 8-1 shows a sample inventory valuation adjustment process.

The need for MRO inventory adjustments is typically identified when cycle counting the inventory. During a cycle count, various components in the inventory are physically counted. When the on-hand quantity varies from what the inventory records show, the valuation adjustment is necessary.

Because adjusting the inventory valuation involves accounting practices, it is always wise to consult the accounting department before making any inventory valuation adjustments. It is likely they will want to develop the appropriate business case, because it may involve company taxes. Once the appropriate business case has been developed, it must be submitted to senior management for approval. If the business case is not properly prepared, it is unlikely to be approved when first presented. The executive management will probably make the team review, reevaluate the circumstances for the adjustment, and resubmit the proposal for approval.

When the proper business case has been constructed, it will be approved. For each line item that needs an adjustment, there will then need to be a reason for the adjustment. There may be some items that have limited shelf life that need to be replaced. However, most inventory valuation adjustments for MRO spares are due to uncontrolled spares loca-

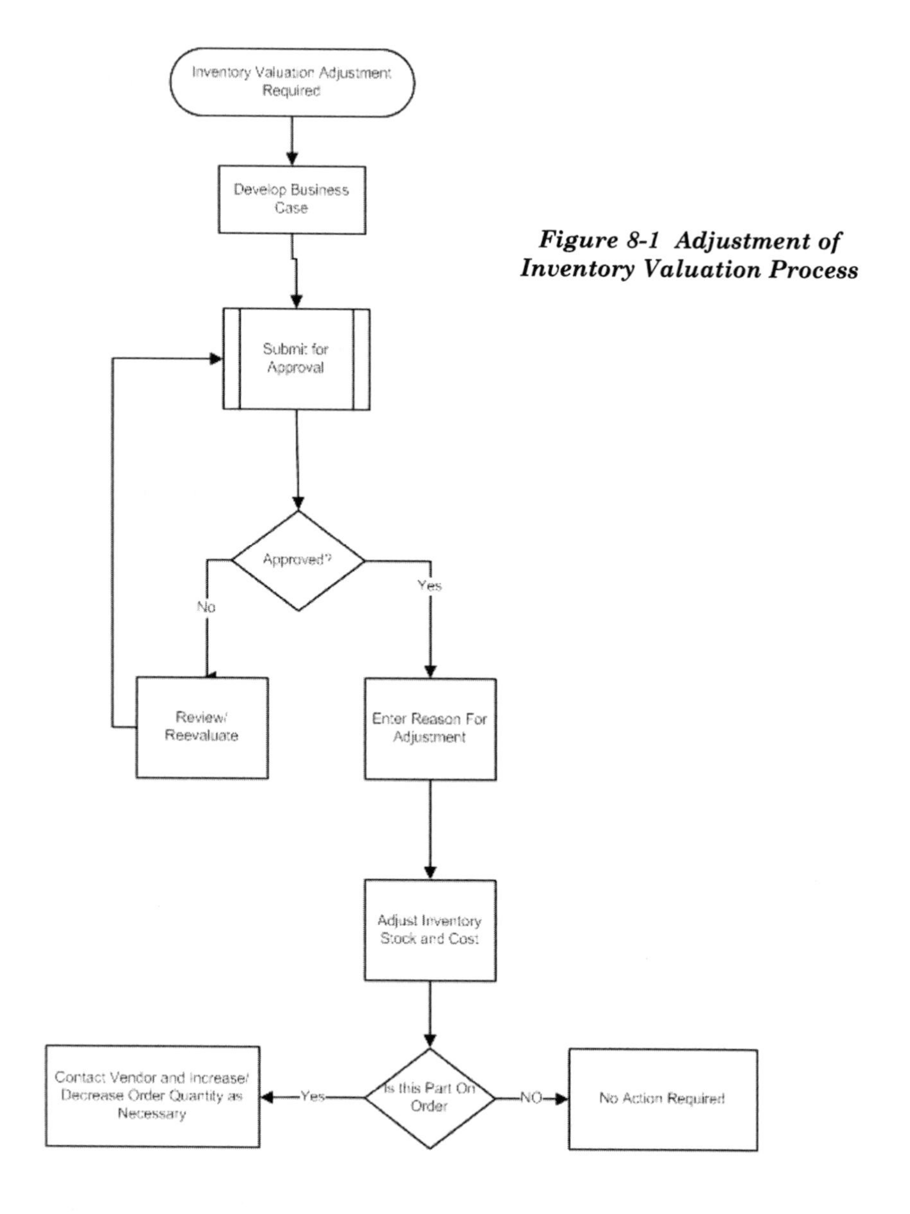

Figure 8-1 Adjustment of Inventory Valuation Process

tions. This simply means the parts usage was not properly tracked and recorded. Tracking the valuation adjustments for MRO spares has cost justified adding storeroom attendants in many locations.

Once the reasons for the adjustment have been clarified, it is simply a matter of updating the on-hand quantities in the computerized inventory system. In some cases, there have been items purchased for projects or contract work, but were not needed. Instead, these items have been put in the storeroom. In this case, the inventory valuation results in adding rather than deleting spare parts from inventory. When this happens, it is necessary to see if there are additional parts already on order. If there are not, then no further action is required. However, if there are spare parts on order, their inventory valuation increases the on-hand quantity. It would then be best to contact the vendor and try to make necessary adjustments to the outstanding order.

Making financial adjustments to an MRO inventory is a transaction that should be taken very seriously. Because federal and state tax laws and corporate liability are involved, the defined corporate process should be closely followed to avoid any possible financial penalties.

Maximizing Investment in MRO Spares

MRO Improvement Initiatives

Once the basics are in place for MRO inventory and purchasing, it is always valuable to reevaluate the processes for possible areas of improvement. The following are areas that can be investigated for possible improvements to maximize the investment in the MRO inventory and purchasing departments.

Maintenance Department-Related Issues

The following areas are ones where the maintenance department can focus and help to reduce MRO inventory costs.

When the maintenance department performs accurate root cause analysis of breakdowns, it reduces the number of repetitive replacements that occur. In many organizations, the maintenance department is constantly changing spare parts. If the root cause analysis is properly performed, the number of repetitive replacements will decrease. In turn, ensuring that the maximum life is achieved for each component or spare part will decrease the MRO budget.

Another process for the maintenance department that can have a dramatic impact on spare parts costing is preventive maintenance. The more standardized that the preventive maintenance procedure is, the less variance there will be in how the preventive maintenance is performed. This performance improvement includes the material usage. If the PM program is standardized, then when the PM is forecast, the materials to be consumed can be forecast as well. Better forecasting will allow the MRO inventory and purchasing personnel ample time to order the materials and have them on hand.

A third area where the maintenance department can have an impact on spare parts cost, is the practice of spare parts hoarding. Initially, a maintenance department may hoard materials when the service level is low from the stores. However, when the stores become more controlled, this practice should stop. If the organization has allowed this practice to become standard, then positive reinforcement should be given to the organization to trust the storeroom personnel and system. When the level of trust is built up, then hoarding of spare parts should stop.

Purchasing-Related Issues

Mature MRO inventory and purchasing departments can save money by reviewing and improving standard MRO practices.

One of the first areas that should be reviewed periodically is the stores organization itself. For example, is the storeroom properly staffed? Are enough attendants on hand to properly order, receive, and stock spare parts? In addition, has there been a change in the geographical structure of the maintenance organization? If so, then a change to the geographical structure of the MRO storeroom, should be investigated. If the storeroom is having a negative impact on maintenance productivity, then a revision to the MRO organization should be made.

Another area that should be reviewed periodically is the work processes in the MRO storeroom and purchasing departments themselves. For example, all recordkeeping systems should be examined to ensure that complete and accurate data is being gathered. In addition, the organization would want to ensure that the right amount of data is being captured. Furthermore, the processes that generate the data should also be examined. The processes themselves should be simple, straightforward, but detailed enough to provide the data necessary to manage the MRO inventory and purchasing function properly. The following are areas that need to be audited periodically:

- Record keeping
- Ordering procedures
- Follow up procedures
- Receiving procedures
- Staging procedures
- Storage procedures
- Issuing procedures

Another area that should be investigated on a regular schedule is the spare parts specifications themselves. This means investigating the materials that spare parts are made of, any new materials that may be available as replacements, and other low-cost alternatives for replacement parts. This focus on specifications insures that, while maintaining quality, the lowest spare parts cost are achieved.

In addition to materials specifications, vendor performance should periodically be reviewed. Each vendor should be evaluated on delivery dates, spare parts cost, and quality of spare parts. While it is not a good practice to constantly change suppliers, the suppliers should know that they are going to be audited periodically. They should then work as hard as possible to make sure the company is receiving the best value for their investment.

Waste and Obsolescence

Another area in which it helps to control the valuation of the inventory is the removal of any unnecessary or obsolete spare parts. When equipment is decommissioned and removed from the plant, all spare parts for the equipment can also be removed from the plant. The exception here is if the spare part is also used on other equipment. Even in this case, the overall maximum and minimum quantities should be reviewed. Removal of these spare parts has a twofold benefit. It removes the value of the spare part from the storeroom, but also frees up space for any new spare parts that will be required.

As with any business process, waste can always be identified. Therefore, with the MRO storeroom and purchasing functions, there should be an aggressive attempt to eliminate all forms of waste, scrap, and spoilage. For example, items with limited shelf life should not be purchased in excessive quantities. Similarly, purchases of parts for equipment that has a short life cycle (which means the equipment will be decommissioned in the near future) should also be closely monitored. Many companies find tremendous opportunities for savings in these areas.

Standardization

The last and possibly largest area where savings efforts can be focused is in the standardization of equipment and material. How many different types of equipment in the plant are used to produce the exact same product or provide the exact same service? Many companies have found tremendous savings in spare parts reductions, simply by standardizing plant equipment. Spare parts should constantly be re-evaluated as to whether or not they can be standardized.

MRO Materials Top Ten Enablers

In addition to the previously mentioned initiatives, benchmarking studies have shown that there are at least ten enablers for successful MRO inventory and purchasing departments. These are highlighted in Figure 9-1.

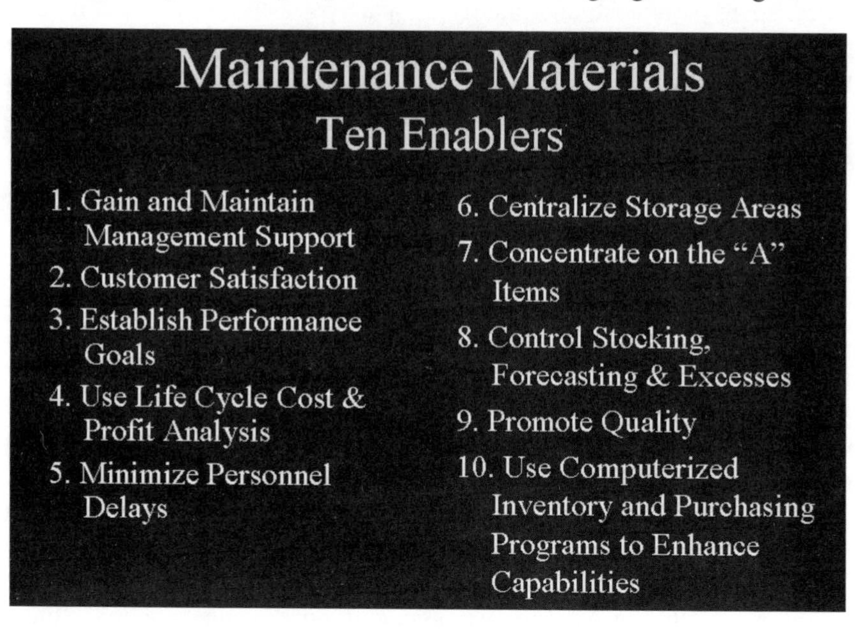

Figure 9-1 Maintenance Materials: Ten Enablers

1. Gain and Maintain Management Support

It is important for MRO inventory and purchasing functions to gain and maintain management support. Without constant support of the persons in charge, the MRO inventory and purchasing departments may seem to be just overhead departments. Management must clearly understand that the inventory and purchasing departments can affect not just the expense side of the ledger, but also the profit side. Although education can gain management support, maintaining management support is driven by proper reporting. Some of the KPIs necessary to maintain management support are highlighted in Chapter 10.

2. Customer Satisfaction

The customer for the MRO storeroom and purchasing function is not just the maintenance department; it is ultimately the shareholders in the company. Although it is important for the MRO storeroom and purchasing function to provide the proper level of service to the maintenance organization, they must do so without incurring excessive expenses that would impact the shareholder's investment. By properly managing the MRO storeroom and purchasing functions, both of these customers can be properly satisfied.

3. Establish Performance Goals

Proper management of the MRO inventory and purchasing departments requires that proper performance goals should be specified. A sampling of these performance goals are listed as follows:

- Stock outs: 3-5%
- Service levels: 95-97%
- Service time
  - (From Requisition to Delivery) — Inventory
  - (From Request to Receipt) — Purchasing
- Number of transactions per shift, day, week
- Number of items per requisition
- Inventory Turnover rate

These types of performance goals should be specified when the new MRO inventory and purchasing processes and policies are put in place. Doing so will help all involved to understand the goals of the processes and policies.

4. Use Life Cycle Cost and Profit Analysis

The goal here is to help reduce expenses. In understanding the life cycle costs of the spare parts, it becomes easier to make decisions when developing part contracts, leasing options, sharing inventories between departments or plants, and developing consignment arrangements. Without these types of detailed financial analysis, an organization may make decisions that lower expenses, but decrease the plant capacity. On the other hand, they can make decisions that will increase the plant capacity, but also increases expenses to the point that it cannot be financially justified. Developing this balance can only be achieved when using life cycle costing and profit analysis.

5. Minimize Personnel Delays

Keep in mind that labor rates are high for most companies. Therefore, labor productivity is very important. The proper MRO inventory and purchasing geographical organizations are important to maintaining or increasing labor productivity. By tracking the request for spare parts, it is easy to evaluate periodically if the correct parts are being stocked at the correct locations.

6. Centralize Storage Areas

The geographical organization for MRO inventory and purchasing is one of the key enablers for best practice organizations. Based on the level of maturity of the maintenance organization, there may be a need to adjust the type of organizational structure for MRO inventory and purchasing. A maintenance organization generally begins in an immature state and it is very difficult to set up delivery systems. This is the reason that most reactive maintenance organizations tend to gravitate toward an area storeroom configuration. However, as the maintenance organization becomes more proactive, and material needs can be anticipated, delivery from a centralized location becomes easier.

If a centralized configuration can be properly utilized, it is more cost-effective—in most cases. This configuration will not only reduce record keeping for the MRO inventory and purchasing departments, but also reduce the MRO inventory and purchasing labor costs. A centralized configuration can also reduce the amount of spare parts that need to be carried in inventory. Note, however, that a centralized configuration utilizing a delivery system is only successful with a very proactive maintenance department.

7. Concentrate on the "A" Items

It is important to concentrate inventory controls on the "A" items in stock. The "A" items make up about 20% of the items in the storeroom, but about 80% of the total cost. These items should be tracked closely, properly stored, and handled properly. Any damaged or missing any items will have a significant impact on the overall MRO inventory and purchasing costs.

8. Control Stocking, Forecasting and Excesses

Controlling the stocking amounts when initiating MRO storeroom and purchasing functions can be difficult. There are vendor recommendations, plants or departments with similar equipment, and employee experiences; can all be used in developing initial stocking policies.

Because the elimination of access is difficult, it is important to monitor stocking levels continuously. When excess materials are eliminated, space, people, and investment costs are all freed up, allowing the remaining resources to focus on caring for these spare parts.

9. Promote Quality

The MRO inventory and purchasing departments want to make sure that the correct parts are received and that they are in good quality condition upon receipt. These departments also want to make sure that the correct parts are being utilized by periodically checking spare part recommendations from the maintenance department.

Quality service from the MRO inventory and purchasing departments can be delivered by following up on maintenance department requests, and any outstanding issues.

10. Use Computerized Inventory and Purchasing Programs to Enhance Capabilities

Computerized inventory and purchasing programs help the department keep accurate records, perform mathematical calculations (such as order point, economic order quantity, average pricing), and track orders. In the past, many MRO inventory and purchasing organizations had to do these tasks manually. With the computerized programs continuing to progress in sophistication, they can be successfully used to manage any size MRO inventory and purchasing organization. A company should not neglect the cost savings and increased service that can be achieved using a computerized inventory and purchasing program.

A concluding point can be made that a good MRO inventory and purchasing organization can only be successful if the total organization has a proactive mentality. Just the same, as maintenance can never satisfy a reactive operations or production department, MRO inventory and purchasing departments can never satisfy a reactive maintenance organization.

10

MRO Inventory and Purchasing Performance Indicators

MRO Inventory and Purchasing

Maintenance materials is the maintenance support function that contributes the most to low maintenance efficiency and effectiveness. It subsequently becomes one of the largest root causes of equipment downtime and capacity losses. MRO spare parts account for an average of 50% of the maintenance budget. It is the second most important function in maintenance (first being preventive maintenance) management. Therefore, it is essential to examine performance indicators that will insure proper management of the stores and procurement functions for maintenance.

1. Inactive Stock Showing No Movement in the Last 12 Months

This indicator is used to find spare parts that are no longer needed. These items may have been purchased as spare parts for equipment that is no longer in the plant. Occasionally items were purchased in large quantities on a one-time basis, possibly for construction or a project, but were not used and have no further use at the plant or facility. Eliminating these items reduces the inventory value and the subsequent holding cost the company must pay.

$$\frac{\text{Inactive Stock Line Items}}{\text{Total Stock Line Items}}$$

As seen by the formula, this indicator measures the number of line items that are inactive divided by the total number of line items carried in the inventory. The percentage shows the opportunity for improvement by eliminating line items from the store inventory.

A second way the indicator can be used is to divide the dollar value of the inactive stock items by the total inventory valuation. This percentage would give the percentage of value that it would be possible to achieve through inventory reduction.

Strengths

This indicator is useful for highlighting opportunities to reduce the overall inventory valuation. In companies where equipment and processes are rapidly outdated by technology changes, this indicator is critical to monitor.

Weaknesses

The weakness of this indicator is that it does not differentiate between a disposable spare and one that is kept on hand due to lead-time and delivery issues. Some equipment spares are kept in stock because the equipment is produced in another country and the lead-time to obtain a spare may be months or even a year. In these instances, keeping the spare is wise, even if it does not move for several years.

Any time this indicator is used to highlight items for possible removal from inventory, careful research should be made to insure that the part does not have a long lead-time or that it is difficult to obtain the spare part. Parts should never be arbitrarily removed from inventory.

2. Stores Annual Turns (Dollar Amount)

This indicator is used to determine the number of times in a year the dollar value of the stores inventory is actually used. Although there are some spare parts that will not be used in a year, many will show numerous turns in a year. This indicator compares the dollar value of the issued items to the total inventory valuation.

This indicator is a widely-accepted benchmark for maintenance stores. The average for companies in the United States is from 1-to-1.2 turns per year. Organizations that practice advanced strategic supplier techniques, are working to raise the number much higher. However, many organizations are still below 1 turn per year.

$$\frac{\text{Total Annual Dollar Amount of Stores Usage}}{\text{Total Inventory Valuation}}$$

Expressed as a decimal number

The measure clearly shows the total annual dollar amount of stores usage as divided by the total dollar value of the inventory. The result is not expressed as a percentage, but rather as a decimal number. It provides the commonly-used number of turns indicator.

Strengths

As a benchmark, this indicator is almost an industry standard. It clearly shows whether a company has too large an inventory (in dollars). The indicator can be used to compare different organizations because the inventory goal is similar for almost all organizations.

Weaknesses

This indicator's greatest weakness is that companies owning a lot of foreign-made equipment will tend to have lower numbers. If the items are arbitrarily removed from inventory to meet some benchmark number, the company may experience a large increase in downtime and a subsequent large decrease in capacity. This indicator must be used carefully.

3. Percentage of Spare Parts Controlled

This indicator highlights the uncontrolled spare parts within a company. In most cases, reactive maintenance organizations tend to develop personal or pirated stores. These spares are never tracked, yet the company paid for them. In many cases, the stores personnel may be re-ordering items that are already in the plant in several untracked locations. The goal is to have all spare parts in controlled stores to insure cost-effective inventory policies.

$$\frac{\text{Total Dollar Value of Maintenance Spare Parts in a Controlled Stores Location}}{\text{Total Inventory on Hand (Estimated Controlled + Uncontrolled)}}$$

As the formula shows, this indicator is the dollar value of all controlled stores items divided by the estimated dollar value of all spare parts. In almost all cases, the total value must be estimated because the actual cost is rarely available. The result is expressed as a percentage. The closer this measure is to 100%, the better are the inventory policies a company is utilizing. In addition, the indicator tends to be lower when the maintenance organization is reactive, driving many expedited orders.

Strengths

The strength of this indicator is that it accurately represents the level of financial control the inventory and procurement departments have over the maintenance spare part value.

Weaknesses

The indicator's greatest weakness is that it is hard to calculate. In most companies, it is difficult (if not impossible) to calculate the value of all of the open storage locations and personal storage locations.

4. Service Level of the Stores

This indicator shows the percentage of time that the stores department was able to fill maintenance requests for spare parts. This indicator is becoming a standard benchmark to compare stores performance. Higher percentages reflect better performances of the stores and purchasing groups in meeting their customers' needs.

$$\frac{\text{Total Number of Orders Filled on Demand}}{\text{Total Number of Orders Requested}}$$

The indicator represents the total number of orders filled on demand divided by the total number of orders requested. The result is expressed as a percentage. The benchmark value for this indicator is between 95% and 97%. Any performance level lower than 95% will contribute to delays in work execution. It will also lead to individuals developing their own storage areas. Values higher than 97% suggest that the store is carrying too many spare parts.

Strengths

This indicator is an accepted measure of stores performance and allows for a fair comparison of stores functions between companies. It has the benefit of being accepted internationally. It has little opportunity for error in its calculation, providing the input data is accurate.

Weaknesses

The largest weakness for this indicator is the opportunity to make an error on the timing of a stock out. The questions that must be asked are: When is the stock out registered? Does it have an impact on the service level? Is it a stock out when the job is being planned but the part is not in stock, or is it when an individual goes to the store room counter and requests the part? When the indicator is calculated at issue time, the percentage should be higher than when at the time of planning. If a company counts the stock out at the time of planning and still has a 95-to-97% service level, they likely have too large an inventory.

5. Stock Outs

Stock outs are the inverse of service level. The stock out indicator represents the number of times the order could not be filled. It too is a widely-accepted indicator. The service level indicator seems to appeal more to customer service oriented organizations, whereas the stock out indicator seems to be used by organizations with a more technical focus. In reality, it doesn't matter which indicator an organization utilizes; they both measure basically the same thing.

$$\frac{\text{Total Number of Items Not Filled on Demand}}{\text{Total Number of Items Requested}}$$

The formula measures the total number of items that were not filled on demand divided by the total number of items requested. The result is then presented as a percentage. The goal is a 3–5% stock out percentage. As with the service level indicator, if the number is too small, then too much inventory is carried. If the number is too large, then delays in work will be experienced.

Strengths

This indicator is an accepted measure of stores performance. It allows for a fair comparison of stores functions between companies. It has the benefit of being accepted internationally. The indicator has little opportunity for error in the calculation, providing accurate data exists.

Weaknesses

The largest weakness for this indicator is the opportunity to make an error on the timing of a stock out. The questions that must be asked are: When is the stock out registered? Is it registered when the job is being planned and the part is not in stock or when an individual goes to the store room counter and requests the part? When the indicator is calculated at issue time, the number should be lower than if it is calculated at the time of planning. If a company counts the stock out at the time of planning and still has a 3–5% stock out level, it is likely they have too large an inventory.

6. Percentage of Rush Purchase Orders

This indicator highlights the amount of reactive ordering that occurs to fill customer orders. A maintenance organization that is reactive drives the ratio of rush purchase orders to a higher level. However, if the maintenance organization is more proactive and the percentage of rush purchase orders is high, then the stores and purchasing function is trying to hold too few spare parts.

$$\frac{\text{Total Number of Rush Purchase Orders}}{\text{Total Number of Purchase Orders}}$$

As the formula indicates, the number of rush purchase orders is divided by the total numbers of purchase orders. The result is expressed as a percentage. Higher percentages indicate a more reactive purchasing function, which will impact the organization with increased expediting costs and a higher rate of downtime. The lower the percentage, the more managed and proactive is the purchasing function, which will allow for planned purchases and consolidated purchase orders, further reducing the overall cost of the inventory.

Strengths

This indicator is useful when examining the cost to process purchase orders. The goal is to keep the percentage as low as possible. However, there are so many factors that impact this indicator, such as customer demand, it should never be used as a single performance indicator for stores and purchasing.

Weaknesses

The weakness of using this indicator is highlighted above—that too many factors outside the control of the stores and purchasing function have an impact on the indicator. For example, if the maintenance organization doesn't plan their work in advance, then the store and purchasing department are forced to make rush orders to meet the demand. This would make the evaluation of their service unfair.

7. Percentage of Single Line Item Purchase Orders

The cost of processing a purchase order ranges from below $50 for smaller companies to over $250 for larger organizations. If the purchase order has only one line item on it, then the cost is additional to the price of that one item. When multiple line items are consolidated on a single purchase order, then the cost of the purchase order per line item (which will increase somewhat, although not in direct proportion) is reduced. This indicator focuses on the percentage of single line item purchase orders. If the number is high, then it is likely that the maintenance department is reactive with many rush requests. If the maintenance department is proactive with forecasted demands, then multiple items can be consolidated on a purchase order, reducing processing costs.

$$\frac{\text{Total Number of Single Line Item Purchase Orders}}{\text{Total Number of Purchase Orders}}$$

The formula divides the total number of single line item purchase orders by the total number of purchase orders. The higher that this percentage is, the more expedited tasks that the purchasing department is performing. The lower the number, the more proactive the purchasing function.

Strengths

This indicator highlights the opportunity to save purchasing processing costs by consolidating line items and reducing the total number of purchase orders processed. It is valuable for determining whether the purchasing function is reactive or proactive.

Weaknesses

This indicator is similar to the previous one in that by trying to meet reactive customer demands, the purchasing function may be unfairly evaluated. This indicator should also not be used as a single indicator for the stores and purchasing function because of the impact of external factors that can not be controlled.

8. Percentage of Maintenance Work Orders Waiting on Parts

This indicator highlights the impact the stores and purchasing function has on the execution of maintenance activities. The higher the percentage is, the more maintenance work that is being held up by the lack of spare parts.

$$\frac{\text{Maintenance Work Orders On Hold Awaiting Parts}}{\text{Total Number of Maintenance Work Orders}}$$

The formula highlights the amount of maintenance activities actually impacted by the lack of parts. The indicator is calculated based on the number of work orders. If the planning function is fully utilized in maintenance, then the hours of work and the cost of the work can also be highlighted. This places yet another perspective on the impact the stores and purchasing function has on maintenance.

Strengths

This indicator shows the maintenance work that is on hold due to parts of the organization other than just the maintenance department. Although it is not a common indicator, it is useful when trying to decide where work execution is a problem.

Weaknesses

If not used carefully, this indicator can be used to place blame. There are dynamic issues impacting this indicator. All must be considered before reaching a decision to take action based on this indicator. These dynamic issues include improper planning of maintenance work orders, excessive amounts of reactive maintenance being performed, and poor organizational disciplines within the inventory and procurement departments.

9. Percentage of Material Costs Charged to a Credit Card

This indicator tracks the usage of credit cards for small purchases in companies today. Although this practice is fairly recent, it is misused in most organizations. Many companies used the credit cards to lower their purchasing costs. Although this goal is a good one, it often compromises the integrity of the material cost data for the equipment life cycle costs. For example, when replacement parts are purchased on a credit card, how is the cost tracked to the equipment history? These may be small items, yet over time, the purchases can add up to a considerable amount. This indicator is monitored to insure there are no abuses of the credit card policy.

$$\frac{\text{Maintenance Material Costs Charged to a Credit Card}}{\text{Total Maintenance Materials Costs}}$$

The indicator is derived by totaling the costs of all items charged to a credit card and dividing that total by the total maintenance material costs. The result is expressed as a percentage. This percentage should be trended to insure that the charges to the credit cards are not excessive or increasing to a high level.

In addition, this indicator should be compared to the total maintenance material costs not charged to a work order. If the maintenance materials are going into a black hole, then it is likely that credit card usage has become excessive.

Strengths

This indicator helps to insure that credit card usage is under control. If the indicator is tracked closely, it can be used to spot abuses or negative trends in the card usage.

Weaknesses

This indicator has no major weaknesses. Credit card use is a practice that needs close control. This indicator is a necessity for any organization utilizing credit cards.

10. Internal Costs to Process a Purchase Order

This indicator is not really calculated by a formula, but rather by tracking the internal costs that are associated with processing a purchase order. This includes the costs involved in processing the purchase order, including the approval levels and time to process. In some organizations, this cost is virtually unknown. Single line item purchase orders are routine because there appears to be no penalty for the practice. Tracking the costs creates an awareness and promotes more multi-line item purchase orders.

Strengths

This indicator is useful for tracking the costs. It helps to insure that the approval policy and processing procedures are cost effective. It should be tracked by all organizations. A monthly posting and trending for a rolling year time period is typical.

Weaknesses

There are no major weaknesses with this indicator. The only time this indicator is not effective is when an organization is not accurately tracking its internal processing costs.

Inventory and Procurement Problems

Figure 10-1 highlights common problems preventing the cost effective optimization of inventory and procurement practices.

Figure 10-1 Stores and Procurement Indicator Tree

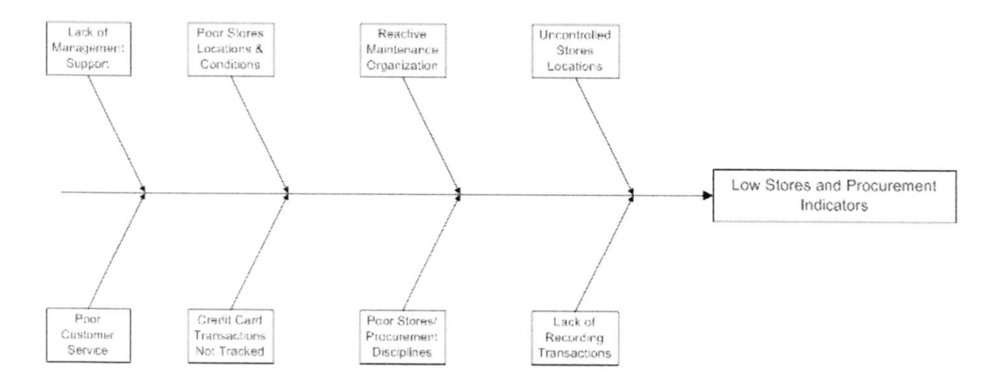

1. Reactive Maintenance Organization

Of all of the items on this list, having a reactive maintenance organization is by far the one that causes the majority of the problems. In fact, it is the key to making the inventory and purchasing functions ineffective. The inventory and purchasing function can not have every spare part that maintenance might need on just a few minutes notice.

An effective preventive maintenance program can reduce the amount of reactive or unplanned work to less than 20% of all maintenance activities. When a large percentage of the maintenance work is performed with a 2-to-4 week notification time, the inventory and purchasing organization has time to respond to the needs of maintenance. Unless the reactiveness of the maintenance department is controlled, there is no possibility of the inventory and purchasing functions performing satisfactorily.

2. Uncontrolled Stores Locations

This problem occurs when the cost of inventory loss is not clearly understood. Many companies today use unsecured storage locations. Although some theft occurs, it is not the major problem. The greater problem occurs when items are used, but are not recorded. The lack of discipline to record the data causes two problems:

- The item is not reordered, creating a stock out and delay next time one is needed.

- The cost of the item is not recorded against the piece of equipment or location where it is used. This creates inaccuracies in the equipment's cost history. It also invalidates any attempt to do life cycle costing of the affected equipment.

Although there have been many attempts to try alternative methods to insure the data is recorded, the only method proven successful over time is to secure and staff the store locations. In addition, the value of the data recording process must be instilled in each employee. Otherwise, there will be constant problems with inventory levels and data analysis.

3. Lack of Recording Transactions

This problem is related to the previous one, but applies even when secured and manned locations exist. The discipline to record data must be instilled in everyone who has a responsibility for issuing, receiving, or returning inventory items. When the value of what is done with the data is not clearly understood, the data collection may seem to be a non-value added function. However, the following scenario should be considered:

If the transaction is not recorded, then the data about stock levels is incorrect. If people who are planning a job and, relying on the data in the system (whether computerized or manual), make a decision based on that data, they will make the wrong decision. They may have a crew of crafts personnel scheduled, a contractor scheduled, and equipment rented. Then when they go to pick up the part, it is missing, even though the system indicates it is there. Meanwhile the costs of the equipment being shut down (when it could have been running) and the resulting lost production costs continue to rise.

The cost of not recording the transactions must be clearly communicated. Once a system is put in place, all employees must discipline themselves to utilize it fully. Anything less will create inventory and purchasing problems.

4. Poor Stores and Procurement Disciplines

This problem occurs when the basics of inventory and purchasing management are not enforced. These include all aspects from initiating an order to receiving, issuing, and recording transactions. The same rule applies in inventory and purchasing management as it does in maintenance: Concentrate on the basics first.

The pattern for inventory management is easy to find—your local auto parts store. The entire operation is well managed; because it has to be profitable, it is controlled. The analogies are abundant. There are self-service areas where you select items and bring them to the counter, and their usage is recorded. There is a secured area, where you cannot personally go back and get your parts. The stores catalog and parts lists are at the counter. What you need is identified and the stores attendant goes back to get the item, brings it to you, and records the transaction. All of your transactions are paid for (charges to an equipment item's account) and typically you go home and perform some work with your purchases.

Many companies today have begun contracting out their stores functions. They claim that there are major financial gains to be realized. It is true that the vendor typically takes the inventory off the company's books, and provides staff to manage the stores, yet this is not the whole picture. The vendor also adds a per transaction charge to each issue. Although this cost is accrued in small increments, it adds up to significant amounts when annualized. In fact, in most cases, the charges amount to more than before, when the company managed its own stores function. This practice is never actually studied and reported in most companies. Because the taxable inventory is gone, the decision to contract out appears to be a good decision.

The decision is often based on what was good for one department or area, but not for the company's overall financials. These decisions should be carefully studied by all departments impacted before they are arbitrarily implemented by one function within the company.

5. Poor Stores Locations And Conditions

This problem is an indication of the priority the inventory and purchasing function has within a company. Ideally, the company will operate stores locations that provide the type of environment that protects and preserves the spare parts while they are in storage. After all, the investment in spare parts for many companies is considerable. Allowing spare parts to deteriorate while in storage is financially wasteful.

A stores location should be able to protect the spare parts from contamination, moisture, and mishandling by unqualified personnel. Yet, there are companies that store major spare parts in open, exposed outside storage areas called Boneyards, where the major components may be exposed to dramatic temperature changes, moisture, and contamination. Then when the spare parts are needed, people wonder why they don't last as long as they should. This problem reflects a lack of knowledge and appreciation for the company's financial investment in the spare parts.

In addition to the condition of the storage, the location of the storage should also be a factor. If the locations are remote, away from the center of the plant, how much time and efforts are required to move the spare parts from the stores to the job? This question should provide a basis for choosing an appropriate location for the stores. If there is considerable time and effort involved in moving the spares from storage to the work area, it will result in lower maintenance productivity, increased downtime, and ultimately reduced equipment capacity.

6. Credit Card Transactions Not Tracked

If the credit card transactions are not tracked correctly, with the charges tracked back to the equipment on which the spare parts are being used, then the cost histories of the equipment items are invalidated. When this occurs, the following results:

- The ability to make cost-effective equipment replacement decisions is forfeited
- The ability to choose the correct replacement based on repair cost is forfeited
- The ability to do life-cycle costing is forfeited

- The ability to decide how much to spend re-engineering equipment to eliminate chronic problems is forfeited

The credit cards may be a cost-effective solution for the inventory and purchasing departments. Yet the impact that credit cards have on all departments' ability to perform their functions should be considered, in depth, before any policies are changed. The ability to eliminate the cost of some clerical work in purchasing is insignificant when compared to the overall impact the decision may have on plant capacity.

If credit cards are utilized, there must be some mechanism to charge the costs accumulated on the cards to the appropriate equipment history. Otherwise, the credit card system should never be implemented for maintenance spare parts.

7. Lack of Management Support

Management must understand the inventory and purchasing functions and support any policy that enhances the overall competitive position of the company. This does not give these departments unlimited ability to set policies. They are service departments, providing services to internal customers. Unless they are responsive and cost effective, the customers may go elsewhere (outsourcing).

However, when it comes to supporting the basics of good inventory and purchasing, management should support any overall cost-effective policies. The proper storage locations, proper staffing, proper procedures, and proper disciplines should always be in place and fully supported. If not, then the stores and procurement function will have no chance to contribute fully to company profitability.

8. Poor Customer Service

This is a major problem in many companies today, where the inventory and purchasing departments try to dictate to their customers policies that conflict with the customers' charter in the company. The inventory and purchasing departments must understand who their customers really are, what their needs are, and how to meet those needs in a manner that is cost effective for the entire company.

Unless customer service is the focus, untold problems develop, with adversarial relations between the customers and the inventory and purchasing groups. It is only when the customers' needs are the focus that an organization can overcome this problem.

GLOSSARY

ABC classification—Method used to categorize inventory into groups based upon certain activity and pricing characteristics.

Activity based costing—Usually refers to costing method that breaks down overhead costs into specific activities (cost drivers) in order to more accurately distribute the costs in product costing.

Actual cost—An inventory costing method used in manufacturing environments; it uses the actual materials costs, machine costs, and labor costs reported against a specific work order to calcu late the cost of the finished item.

Advanced shipment notification—Advanced shipment notifications (ASNs) are used to notify a customer of a shipment. ASNs will often include PO numbers; SKU numbers; lot numbers; quantity, pallet, or container numbers; carton numbers.

Aisle—Any passageway within a storage area.

Allocations—Allocations in inventory management refer to actual demand created by sales orders or work orders against a specific spare part. The terminology and the actual processing that controls allocations will vary from one software system to another.

- A **soft allocation** is an aggregate quantity of demand against a specific item in a specific facility. I have heard standard allocations referred to as normal allocations, soft allocations, soft commitments, regular allocations. Standard allocations do not specify that specific units will go to specific orders.
- A **hard allocation** is an allocation against specific equipment or assets within a facility, such as an allocation against a specific location, lot, or serial number. Firm allocations are also referred to as specific allocations, frozen allocations, hard allocations, hard commitments, holds, reserved inventory. Standard allocations simply show that there is demand whereas firm allocations reserve or hold the inventory for the specific order designated.

Approved brand or equivalent specification—A specification referencing a certain brand and model of a spare part that meets the quality and performances required. This type of specification may allow bidding of other manufacturers' brands which comply with the standards called for. Also known as a "Qualified Part."

Approved brand specification—A specification referencing a brand and model or certain manufacturer's spare part. This specification does not allow equivalent brands.

As Is—An indication or notice that the provider of goods will not be responsible for the condition or performance if the purchaser accepts them.

ASP (Application service provider)—A service where the software licenses are owned by the ASP and reside on their system while the client rents the rights to use the software. The ASP may be the software manufacturer or a third-party business.

Authorized price list—A list of spare parts or services resulting from a contract which provides agreed-upon prices and the necessary information to place orders.

Automated data collection—Systems of hardware and software used to process transactions in warehouses and manufacturing operations. Data collection systems may consist of fixed terminals, portable terminals and computers, radio frequency (RF) terminals, and various types of bar code scanners.

Available—The status of a spare part as it relates to its ability to be reserved or issued. Availability calculations are used to determine this status. Availability calculations vary from system to system but basically subtract any current allocations of holds on inventory from the current on-hand balance.

Average cost—Inventory costing method that recalculates an item's cost at each receipt by averaging the actual cost of the receipt with the cost of the current inventory.

Back-order—An unfilled request for issue of warehouse stock. The term requires the generation of a purchase order for stock replenishment if not already ordered. Back-order filling is a matter for policy statement.

Bay—Designated area within a section of a storage area, or a shop, outlined by marking on columns, posts, or floor. Usually, a specific area within a section, such as 20' x 20' squares.

Bill of lading—A document by which a transportation line acknowledges receipt of spare parts or materials and contracts for its movement.

Bill of material—(BOM) lists the spare parts carried in stock for a specific piece of equipment. Multilevel BOMs also show sub assemblies and their components.

Blanket order—A type of purchase order that commits to purchase a specific quantity over a specific period of time, but does not necessarily provide specific dates for shipments

Carousel—Type of material handling equipment generally used for free issue items

Carrying cost—Also called holding cost, carrying cost is the cost associated with having inventory on hand. It is made up primarily of the costs associated with the inventory investment and storage cost. Carrying cost is represented as the total annual cost divided by the on-hand inventory unit quantity.

Chargeback—A financial penalty placed against a supplier by the company when a shipment does not meet the agreed upon terms and conditions.

Commodity—The term used to describe classifications of inventory. Therefore, "commodity codes" are used to distinguish groups of inventory items to be used for reporting and analysis.

Competitive quotations—A purchasing method used to obtain competitive pricing for spare parts when the anticipated cost is less than the amount required for formal or competitive sealed bids. This method is used only for small purchasers and should be documented or recorded by written statements from the suppliers.

Consignment inventory—Spare parts inventory that is in the possession of the customer, but is still owned by the supplier.

Container—Anything designed to hold (contain) materials for storage or transport,

Contingency—An allowance made to provide for or protect desirable conditions in the future. Such conditions may be threatened by causes or events unforeseeable in the future.

Contract administration—The management and monitoring of legal agreements to ascertain that the contractor's commitments to the company are being fulfilled according to contract.

Contract warehouse—A business that handles shipping, receiving, and storage of products on a contract basis.

Cost of money—The amount of interest that would be earned if the dollar value of inventory were invested at the company's current investments earning rate.

Critical stock—A stock item that must be maintained in inventory, though little used, to respond to the equipment maintenance needs. These parts are required to keep equipment or assets operating.

Current cost—Inventory costing method that applies the cost of the most recent receipt to all inventory of a specific item.

Cycle count—The process of regularly scheduled inventory counts that "cycles" through the inventory. This insures all inventory items are counted and reconciled in the course of a complete fiscal year.

Delivery terms—A contractual designation of location of delivery, the time of delivery, and shipping costs.

Demand—The need for a specific item in a specific quantity.

Economic order quantity—Result of a calculation that determines the most cost effective quantity to order. The formula specifies the point at which the combination of order cost and carrying cost is the least.

Emergency stock—The quantity of a part that must be maintained on hand at all times to provide for initial response to an unplanned catastrophic event.

Excess Inventory—Inventory quantities above the specified maximum quantity.

Expediting—A rush shipment or follow up on orders placed to insure timely shipment and delivery. Usually the most expensive delivery system.

FIFO—First-in-first-out. In warehousing, describes the method of rotating inventory to used oldest product first.

Fixed-price contract—A contract requiring that prices remain firm. During the term of such contracts the contractor must absorb any increases which would reduce their profit

Fulfillment—The activity of processing customer shipments.

Functional or performance specification—A specification which places emphasis on describing a result or capability to be accomplished by a spare part or service. A method of inspection or testing may be included.

Guidance systems—Used to guide automated guided vehicles through plants, guide lift trucks in very-narrow-aisle storage areas.

Independent demand—Demand generated from forecasts, customer orders, or service parts.

Industrial truck—Vehicles used for industrial purposes. Generally used to transport materials and personnel within industrial facilities. Lift trucks (forklifts) are the most well known type of industrial truck.

Inspection report—To inform the purchasing department that an exami nation or testing of received spare parts has taken place. Such a report would further inform purchasing of the quality or condition of such goods.

Inventory—Any quantifiable item that you can handle, buy, sell, store, issue, produce, or track can be considered inventory.

Inventory error—Net Error (net book to physical difference), it is expressed as dollars using the sum of all inventory loss and all inventory gains.

Inventory management—The direction and control of activities with the purpose of getting the right inventory in the right place at the right time in the right quantity in the right form at the right cost.

Inventory turnover (Turnover, Turnover Rate, Stock Turn)-The mathematical determination of the number of times there has been a complete replacement of inventory stock in a years time. Usually determined on a dollar-value basis.

Invoice (bill) -A list of charges or costs presented by a vendor to a pur chaser, usually enumerating the items furnished, their unit and total costs, and a statement

Kanban—Used as part of a Just-In-Time production operation where components and sub-assemblies are produced based upon notification of demand from a subsequent operation. Historically, Kanban has been a physical notification such as a card (kanban cards) or even an empty hopper or tote sent up the line to the previous operation.

Lead time—Amount of time required for an item to be available for use from the time it is ordered. Lead time should include purchase order processing time, vendor processing time, in transit time, receiving, inspection, and any internal delivery times.

Lean manufacturing—Alternate term used to describe the philosophies and techniques associated with Just-in-time (JIT) manufacturing and the Toyota Production System .

Legacy system—Implies a business computer/information system that is old or outdated.

Life cycle costing—The total cost of owning a spare part or piece of equipment for a designated period of time. The time period is usually based upon need or the expected life of the spare part or piece of equipment. This is an engineering method that takes into account all expenses of ownership such as: purchase price, maintenance, operating costs, decommissioning costs, and remaining value at the end of ownership.

LIFO, Last-in-first-out—In warehousing, describes the method for using the newest inventory first. In accounting, it's a term used to describe an inventory costing method.

Lift truck—Vehicles used to lift, move, stack, rack, or otherwise manipulate loads.

Line item—A single detail record. The term *line item* is most commonly used to describe the detail (each line that reflects an item and a quantity) on purchase orders. For example, if an order is placed for 10 red pens, 20 black pens, and 50 green pens, this equates to an order with three line items.

Locator system—Inventory-tracking systems that allow you to assign locations to your inventory to facilitate greater tracking and the ability to store product randomly.

Lockout / Tagout—The process of disabling (lockout) and identifying (tagout) equipment and energy sources during maintenance or service to prevent injury of personnel from an unexpected start-up or power up.

Maintenance, repair, and overhaul inventory (MRO)—Inventory used to maintain equipment as well as miscellaneous supplies such as office cleaning supplies.

Manufacturing execution system (MES)—Software systems designed to integrate with enterprise systems to enhance the shop-floor-control functionality that is usually inadequate in ERP systems. MES provides for shop floor scheduling, production, and labor reporting; integration with computerized manufacturing systems such as automatic data collection; and computerized machinery.

Mezzanine—A tiered structure within a building used to provide store room access to various levels. Mezzanines can be free-standing structures supported by posts and trusses, or can be a series of walkways supported by storage equipment (rack-supported mezzanine).

Minimum inventory—The amount of stock on hand that has been designated as safety stock. When this quantity is reached, a reorder is processed.

Min-max—A basic inventory system in which a minimum quantity and maximum quantity are set for an item. When the quantity drops below minimum, orders are place to bring the quantity on hand back to the maximum.

Obsolete Inventory—Spare parts that are no longer usable for their intended purpose through expiration, contamination, or change of need.

Optional replenishment—The action of ordering or producing up to the max in a min-max system even though inventory has not reached the Min. May be used to avoid downtime on machines, etc.

Order cost—Also known as purchase cost or set up cost, order cost is the sum of the fixed costs that are incurred each time an item is ordered. These costs are not associated with the quantity ordered but primarily with physical activities required to process the order.

Order level—The level of stock of any spare part at which an order is initiated for the restocking of that part.

Overhead—Includes all of the factors other than direct labor and materials included in the cost of goods sold. This figure is usually expressed as a percentage of direct labor cost, a dollar amount per production unit, and several other ways.

Packing List—A document that itemizes in detail the contents of a particular package or shipment.

Pallet —A portable platform designed to allow a forklift or pallet jack to lift, move, and store various loads. Most pallets are made from wood, but pallets are also made from plastic, steel, and even paper-based materials

Paperless—When referring to processing in MRO (paperless picking, paperless receiving) or on the shop floor, paperless generally suggests that the direction of tasks and execution of transactions are conducted electronically, without the use of paper documents.

Performance bond—A guarantee submitted by a contractor, certifying that if the contractor is unable to fulfill the obligation, the specified amount will be paid to the purchaser to compensate any loss.

Performance record—A documentation of the contractor's ability to comply with the requirements of a contract during the term.

Performance specifications—A specification that places emphasis on describing a capability or result to be accomplished with a spare part or service. A testing or inspection may be included.

Physical inventory—Refers to the process of counting all inventory in a warehouse or plant. MRO operations are usually shut down during a physical inventory.

Price agreement—A price decided upon between the company and the vendor(s).

Procurement cycle—The entire cycle of purchasing functions and duties which occur during acquisition of spare parts.

Purchase order—A document used to approve, track, and process purchased items. A purchase order is used to communicate a purchase to a supplier. It is also used as an authorization to purchase. A purchase order will state quantities, costs, and delivery dates. The purchase order is also used to process and track receipts and supplier invoices/payments associated with the purchase.

Quality assurance—A program planned to provide that spare parts purchased may be inspected and/or tested so that compliance with specifications may be determined.

Quality control—A monitoring of a manufacturing process that determines the level of quality of the goods that are being supplied.

Quantity on hand—Also known as on-hand quantity, in stock, store quantity Quantity on hand describes the actual physical inventory in the possession of the business. When inventory is received or produced, it is added to quantity on hand. When inventory is sold or consumed, it is removed from quantity on hand.

Quantity on order—Includes quantity on open purchase orders or manufacturing orders. May or may not include quantities on transfer orders from other branches.

Quantity in transit—In multi-branch environments, quantity in transit reflects the quantity that has been shipped from one branch/facility to another branch/facility, but has not yet been received by that branch/facility. In operations that use advanced tracking of receipts, it may reflect quantities that have been shipped by outside vendors, but not yet received.

Quantity allocated—Also known as committed quantity, commitments, or allocations. Quantity allocated is the quantity that is on current open sales orders or production orders (as components), and may be relative to a specific time period. Also see **Allocations.**

Quantity available—The result of a calculation that takes quantity on hand and reduces it by allocations (e.g., sales orders, manufacturing orders). Quantity available may or may not be date specific and therefore take into account future receipts. Quantity available calculations are sometimes very complicated and vary from one software product to another.

Radio frequency—In warehousing, refers to the portable data collection devices that use radio frequency (RF) to transmit data to host system.

Real-time locator system—Real-time locator system (RTLS) uses RFID technology that provides the objects they are attached to the ability to transmit their current location. System requires some type of RFID tag to be attached to each object that needs to be tracked, and RF transmitters/receivers located throughout the facility to determine the location and send information to computerized tracking system.

Receiving report—Form used by the receiving unit to inform others of the receipt of spare parts purchased.

Reorder point—The inventory level set to trigger reorder of a specific item. Reorder point is generally calculated as the expected usage (demand) during the lead time plus safety stock.

- **Fixed reorder point** implies the reorder point is a static number plugged into the system.
- **Dynamic reorder point** implies there is some system logic calculating the order point. Generally this would be comparing current inventory to the forecasted demand during the lead time plus safety stock.

Requisition—Internal document by which maintenance (or another department) notifies the purchasing department to initiate a pur chase. May also be a form used by the maintenance department to obtain supplies from a storeroom or warehouse.

RFID—Radio frequency identification. Refers to devices attached to an object that transmit data to an RFID receiver.

Rolling average inventory—An average inventory valuation based on the immediate past 12-month period

Routing—Used in conjunction with the bill of material in manufacturing operations. Although the BOM contains the material requirements, the routing will contain the specific steps required to produce the finished items. Each step in the routing is called an operation; each operation generally consists of machine and labor requirements.

Safety lead time—A way to represent the safety stock as a number of days demand.

Safety stock—The level of stock, over and above the expected usage between the time a replenishment order is processed and replenishment actually occurs, that is held in reserve to try to prevent stock-out, should there be a delay in delivery of stock by the vendor.

Salvage—Property or equipment which has served the useful life, but still has value as a source for parts or scrap.

Scheduled purchases—A regulated bidding to be carried out at predetermined intervals to coincide with the volume acquisition needs of user agencies.

Scrap—Spare parts that are deemed worthless to the owner and are only valuable to the extent they can be recycled.

Sealed bid—A bid submitted as a sealed document, by a prescribed time, by a specific vendor. The contents of the bid will not be known to others prior to the opening of all bids.

Seasonality—Fluctuations in demand that repeat with the same pattern over equivalent time periods.

Service factor—Factor used as a multiplier with the *Standard Deviation* to calculate a specific quantity to meet the specified service level.

Shelves—Board fixed horizontally and supported by a frame or uprights. May be made of metal or wood. Shelves may be fixed or adjustable. Used for small stores.

Shrinkage—“Vanished” spare parts losses and losses by issue error.

Skid—A portable platform designed to allow a forklift, pallet jack, or other material handling equipment to lift, move, and store various loads.

SKU (Stock Keeping Unit)—Referring to a specific item in a specific unit of measure.

Sole source—A purchase made without issuance of competitive bids for a commodity that is known to be available from only one source.

Speech-based technology—Also known as voice technology, it is actually composed of two technologies:

- **Voice directed,** which converts computer data into audible commands
- **Speech recognition,** which allows user voice input to be converted into data.

Standard cost—Inventory costing method used in manufacturing environments. It combines the materials costs in the bill of materials with the labor costs (based on standard labor hours and rates per operation) and machine costs in the routing to calculate the cost of the finished or semi-finished item.

Stock-out—The condition existing when a supply requisition cannot be filled from stock.

Stock-out rate—The number of stock-outs per hundred line items picked.

Storeroom—A secure place for storage of things. A storeroom may be a designated, separate secure area within a warehouse or a designated storage room in a workplace, and may contain ware house stock or end-use items.

Supplies—All items that are consumable. Generally, these would be items with a short life, usually being completely utilized (consumed) once issued.

Transportation management system—Category of operations software that may include products for shipment manifesting, rate shopping, routing, fleet management, yard management, carrier management, and freight cost management.

Turnaround time—The length of time for completing a process usually expressed as an average (usually number of days), but can be determined for individual items (e.g., time from placing requisition to issue of goods, or time from requisition to issuance of a purchase order).

Unit of measure (U/M)—Describes how the quantity of an item is tracked in your inventory system. The most common unit of measure is "each" (EA), which simply means that each individual item is considered one unit. An item that uses "cases" (CA or CS) as the unit of measure would be tracked by the number of cases rather than by the actual piece quantity. Other examples of units of measure would include pallets (PL), pounds (LB), ounces (OZ), linear feet (LF), square feet (SF), cubic feet (CF), gallons , thousands, hundreds, pairs, and dozens.

Unit-of-measure conversions—A unit-of-measure conversion is needed when working with multiple units of measure. For example, if you purchased an item in cases (meaning that your purchase order stated a number of cases rather than a number of pieces) and then stocked the item in "each," you would require a con version to allow your system to calculate how many "each" are represented by a quantity of cases. This way, when you received the cases, your system would automatically convert the case quantity into an each quantity.

Vendor File—A file maintained that lists vendors. The file should contain all information pertinent to the vendor, i.e., application information, commodities supplied, and performance record.

Vendor-managed inventory (VMI)—Phrase used to describe the process of a supplier managing the inventory levels and purchases of the materials they supply. This process can be very low tech, such as an office or maintenance products supplier coming into your facility once per week to visually check stock levels and place a re-supply order, or high tech, such as an electronic component supplier having remote access to your inventory management and MRP system, then producing and automatically shipping to meet your production schedule. Vendor-managed inventory reduces internal costs associated with planning and procuring materials and enables vendors to better manage their inventory through higher visibility to the supply chain. Vendor-managed inventory may be owned by the vendor (consignment inventory) or the customer.

Wire-guided—Term used to describe vehicles that use a wire embedded in the floor to guide the vehicles.

Appendix – A

Economic Order Points, Reorder Points, and Safety Stock for MRO Inventory

MRO Inventory Systems

MRO spare part inventories are supplies that are used in manufacturing, but never become part of the finished product. Examples of these inventories include equipment repair parts, lubricants, hand tools, disinfectants, and other supplies. The objectives of the MRO inventory control are to achieve a satisfactory level of service for the maintenance organization, while keeping the MRO inventory costs as low as possible. Factors that influence the levels of MRO inventory include the service level required, and the cost of ordering and then holding the inventory in storage.

The basic purpose of MRO inventory management is to specify what spare parts should be ordered, when they should be ordered, and the number of spare parts in the order. To accomplish this, effective MRO inventory management policies and processes need to be in place. This means there needs to be a computerized system to keep track of inventory. With the inventory properly tracked, a reliable forecast of demand use for each spare part can easily be obtained. By tracking order and delivery dates, the MRO organization will develop accurate lead times. In addition, the MRO inventory management system should also track reasonable estimates of holding costs, ordering costs, and stock out costs. Although this information varies based on the type of spare part, a classification system should be in place to allow for these variances. This system, typically known as an ABC classification system, will be explained at the end of this appendix.

Key Terms

In order to establish an effective MRO inventory system, certain terms must be understood. Figure A-1 highlights these terms. Lead-time is the time between placing the order and actual receipts of the spare parts

7 Key MRO Inventory Terms

- *Lead time*: time interval between ordering and receiving the order
- *Reorder point* - When the quantity on hand of an item drops to this level, the item is reordered
- *Safety stock* - Stock that is held in excess of expected usage due to variable usage rate and/or lead time
- *Service level* - Probability that an order can be filled on demand
- *Holding (carrying) costs*: cost to carry an item in inventory for a length of time, usually a year
- *Ordering costs*: costs of ordering and receiving inventory
- *Stock out costs*: costs incurred when an order cannot be filled

Figure A-1 7 key MRO Inventory Tearms

from the vendor. It is an important factor because longer lead times lead to higher inventory levels. In order to prevent a stock out, the inventory level must be inflated to compensate for the part usage while waiting for the delivery from the vendor. Conversely, shorter lead times can lead to lower inventory levels because the part will be delivered before the inventory level drops to zero.

The reorder point is the stock level at which an item is reordered. It is heavily influenced by both the item usage and the lead time. If the usage is rapid and the lead time is long, the reorder point will be high to prevent a stock out. Conversely, if the part usage is slow and the lead time is short, then the reorder point will be at a lower level. Safety stock is the spare part quantity that is kept above the expected usage (during the order cycle) due to historically variable usage, or lead time. Safety stock is used to insure higher service levels. The service level is the probability that an order can be filled on demand. Service levels for MRO spare parts should average 95–97%.

Service Levels

MRO organizations must balance their service levels with the cost of holding an inventory. For example, holding or carrying costs are annualized cost to store spare parts. These costs include lighting, storage space,

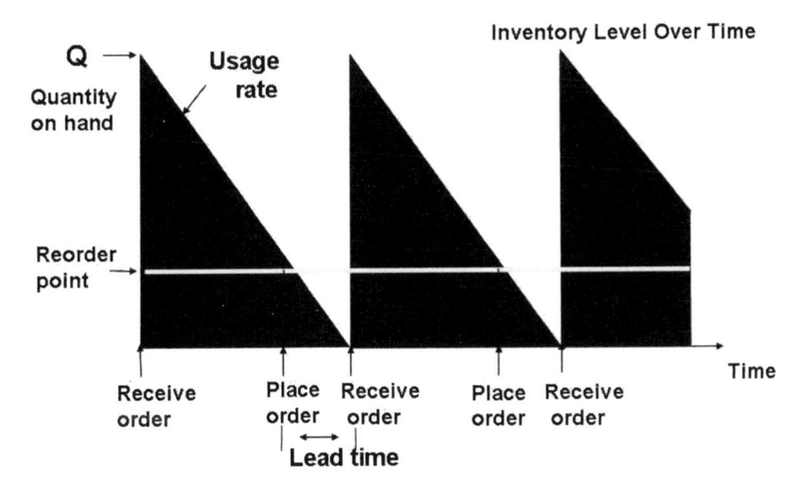

Figure A-2 The MRO Inventory Cycle

and storeroom labor. Ordering costs, which include the cost of processing a purchase requisition, converting it to a purchase order, and receiving the spare parts, are a factor. Stock-out costs are the lost opportunity cost for production while waiting for delivery of spare parts when equipment is idle. These costs together can run into hundreds, if not thousands of dollars per hour. They drive organizations to have high service levels from MRO storerooms.

The MRO Inventory Cycle

To understand the relationship between service levels and inventory costs, consider Figure A-2. This is the MRO inventory cycle. It begins with a quantity on hand, represented by the letter Q. As items are used from inventory, the usage rate develops a geometrical profile. The quantity on hand will diminish at this rate until it reaches the reorder point. When this point is reached, an order will be placed. The time between the placement of the order and the receipt of the order is the lead time. When the order is received, the on hand quantity is typically replenished. This cycle continues to repeat itself over time.

As long as the usage of the spare part is consistent, the process can be predicted and the reorder point set, with an automatic reorder cycle. However, if the spare part usage is inconsistent and unpredictable, then the reorder point will need to be periodically reevaluated. The reorder cycle may be a manual process in which items at or below their reorder point are flagged for review by the MRO storeroom personnel. The order is then placed manually after the review is completed and the purchase approved.

Holding Inventories

At first look, this cycle seems to be a simple one. Yet it becomes more complex when financial considerations are included. For example, what is the most cost-effective quantity to order when placing the order? What is the impact of the predictability of the usage rate on the time between orders? To answer these questions properly, a deeper look at inventory costs is in order.

Figure A-3 highlights four considerations for holding MRO inventories. The first is to provide a quick response to reactive work. Simply stated, this means that when equipment breaks down, spare parts need to be available to repair it. The timeliness of obtaining the spare parts affects the Mean Time To Repair (MTTR) indicator for the maintenance organization. The more reactive a maintenance organization is, the greater the number of spare parts they need to keep this indicator as low as possible. Reducing this indicator to as low a level as possible will drive the organization to keep excessive inventories.

There are at least four reasons for holding MRO inventories:

- To provide quick response to reactive work
- To provide the ability to timely schedule normal backlog work
- To increase the maintenance department's efficiency
- To reduce the amount of downtime incurred while making a repair to equipment

Figure A-3 Holding Inventory

In addition to reactive work, spare parts are required for routine work. For the routine work to be scheduled from the backlog, a certain number of spare parts need to be on hand for the work orders scheduled each week. Balancing the on-hand spare parts with the work orders to be performed in the next two-to-four weeks is part of the MRO inventory department's responsibility. If the maintenance department is to be effective and efficient in performing their work activities, then a certain level of spare parts needs to be on hand. If the maintenance department personnel start jobs, and then have to stop working on them because of the lack of spare parts, their efficiency is impacted.

The organization also needs to reduce the amount of downtime incurred while making unexpected repairs to the equipment. Equipment downtime is costly. Therefore, it is critical to have the spare parts necessary to shorten the mean time to repair (MTTR). If delays impact the delivery of production parts to customers, the profitability of the entire company could be impacted.

MRO Costs

If these factors were the only financial considerations for establishing inventory levels, every company would have enormous MRO storerooms. However, cost constraints must also be considered. Four classifications of these MRO-related costs are listed in Figure A-4.

Item cost is the purchase price of the individual spare part. Total item cost equals the spare part cost times the total quantity on hand or on order. Understanding this cost is important because the total of all item costs equal the inventory valuation for the entire MRO storeroom. Knowing the inventory valuation is critical for financial measures that are typically monitored by the company and usually reported to the Internal

- There are at least 4 classes of MRO costs:
  - Item cost
  - Holding cost
  - Ordering costs
  - Stock out costs

Figure A-4 MRO Related Costs

Revenue Service. Many companies track the total inventory valuation on a monthly basis to insure that the MRO inventory level is not becoming excessive.

Carrying or holding costs are typically charged as a percentage of the dollar value. For example, if the carrying cost rate is 30% (typical for companies in the United States), the carrying or holding costs will be $.30 for every one-dollar worth of the inventory value per year. For a company with $10 million in MRO inventory value, the annual MRO holding costs could be up to $3 million per year. What makes up these costs? They are comprised of the cost of capital, the cost of storage, and obsolescence. The cost of capital is the penalty paid for having the company's capital tied up in the spare part rather than being invested and earning a return (such as interest or dividend) somewhere else. The cost of storage includes the cost of the facility floor space, the insurance carried on the MRO inventory, depreciation of the inventory, and the taxes that must be paid on the inventory. The cost of obsolescence include spares that have reached the end of their shelf life, spares that have been damaged in storage, and spares that are missing due to being lost or stolen.

Ordering costs are the costs incurred while processing a purchase requisition or a purchase order for inventory replenishment. This cost includes initiating, approving, and forwarding a purchase order. The more individuals who are required to approve a purchase requisition or purchase order, the greater the ordering cost will be. Ordering costs may include a charge for expediting the purchase order, as well as transportation and receiving costs. Ordering cost are typically independent of the size of the order, and are typically charged per purchase order. This cost can range from $50 per order in smaller companies to $250 (or more) per order in larger companies. In extreme cases, organizations have had purchase order processing costs of over $1,000. Ordering costs are the primary driver for consolidation of purchase orders. If multiple line items are placed on a purchase order, there is a reduction in the per line item rate of processing. Although the reduction does not have a one-to-one relationship, it can be significant. The potential for savings is the reason that MRO organizations try to become more proactive in their demand forecasting, because it will make purchase order consolidation simpler.

Stock out costs (or downtime costs) are related to the spare part not being available. They include items such as wasted labor from both maintenance and operations; the cost of lost production while the spare part is being ordered, shipped, and delivered; penalty costs for late deliveries, overtime that was worked to make up the lost production; and increased utility costs for using the equipment longer than necessary to meet the production goals. These costs can quickly add up to tens of thousands of dollars in unnecessary costs.

Economic Order Quantity

Attempting to balance all of these costs has led to developing the economic order quantity (EOQ) for spare parts. The EOQ is the key to controlling MRO inventories by focusing on the number of parts to order and when to place the order. The economic order quantity is the order size that minimizes the total annual cost to the company, factoring in all cost considerations. Several basic assumptions are made when using the economic order quantity model. These are:

1. Only one spare part is considered at a time.
2. A historical average annual demand is known.
3. Usage is consistent throughout the year.
4. Lead time does not vary.
5. Each order is received in a single delivery.
6. There are no quantity discounts.

The formula to calculate the economic order quantity is highlighted in Figure A-5. The formula shows that the total cost is equal to the annual carrying costs plus the annual ordering costs. The detailed formula shows that the order size in units (Q) divided by two, multiplied by the

$$\text{Total cost} = \text{Annual carrying cost} + \text{Annual ordering cost}$$

$$TC = \frac{Q}{2}H + \frac{D}{Q}S$$

Q = Order size – in units
H = The cost of holding one unit per year
D = The annual demand – in units
S = The cost of setting up an individual order

Figure A-5 Total MRO Cost

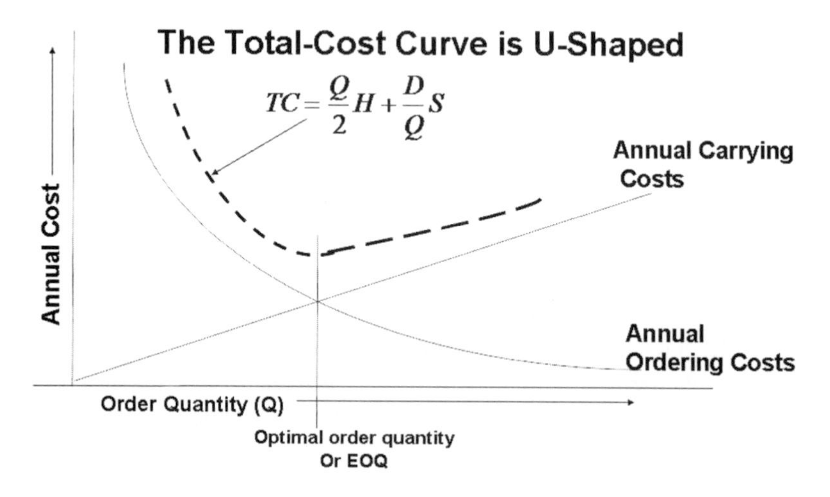

Figure A-6 Total MORO Cost Minimization

holding cost of one unit per year will give the annual carrying costs. The annual ordering costs are determined by the annual demand (D) divided by the order size, both in number of spare parts or units, times the cost of setting up an individual order. The balance of these two costs provides the total cost.

In theory, the only two factors that can impact the calculation are the order size (Q) and the annual demand (D). The holding cost and the cost of setting up an order should remain constant during the calculation. By adjusting the order size and the annual demand, the optimum economic order quantity can be achieved.

Figure A- 6 provides a closer look. Here, the total cost curve is the curved dashed line. The optimal order quantity is the lowest point on the total cost curve. The curved dashed line represents the sum of the annual carrying or holding costs and the annual ordering costs. As the annual carrying costs increase, which indicates carrying a higher inventory levels, fewer orders will need to be placed, which lowers the annual ordering costs. Conversely, low annual carrying costs indicate a constant reorder situation, which will then drive the annual ordering costs high. The optimal or economic order quantity is the lowest point on the combination of the two lines—the lowest total cost on the total cost curve.

Suppose the order size (Q) is 10 units, a unit is valued at $1,000, and the holding cost percentage is 30%. The annual carrying cost is [(10

Safety stock reduces risk of stock outs during lead time

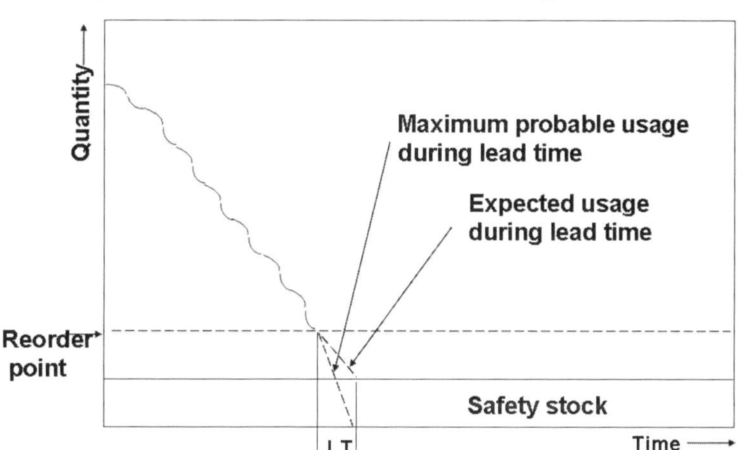

Figure A-7 Safety Stock

x 1,000 x .3)/2] = $1,500. If the annual demand is 1200 units and the cost of setting up the order is $100, the annual ordering costs is [(1200/10) x 100] = $12,000. This makes the total cost for the current policy ($1,500 + $12,000) = $13,500. However, is this the lowest total cost achievable? A spreadsheet can be constructed that shows the annual carrying cost and the annual ordering costs. From this spreadsheet, the lowest total cost can be determined and, in turn, the economic order quantity adjusted for the lowest total cost. This method allows an organization to achieve the lowest total cost when establishing the economic order quantity.

The factors that determine the economic order quantity are the quantity of part usage, the lead time, variability in either demand or lead time, the probability of a stock out, and the related costs. Once the data is available to determine these factors, the optimum reorder point and the economic order quantity can be determined.

Reorder Points

Establishing a reorder point, or the point when an order is initiated, is a matter of looking at the lead time and the usage. If the goal is to achieve zero stock outs, the reorder point should be established at the level when the usage will drive the on-hand quantity to zero at the time the new order is received. This calculation is straightforward, provided all of the data is available and is accurate.

Safety Stock

One additional factor has an impact on the reorder point—safety stock. As mentioned previously, safety stock is stock held in excess of the expected usage, due to variable usage rates or variability in lead-time. The level of safety stock is based on many criteria. The most common and simple approach begins by calculating the normal reorder point for a baseline. With the baseline established, the maximum probable usage during lead time should be used (instead of the expected usage during lead time) to determine the level of safety stock. Figure A-7 shows the impact of safety stock on the reorder point. As the figure indicates, the slope of the maximum probable usage during lead time is greater than the slope of the expected usage during lead-time. If safety stock is not added to the reorder point, there will be a stock out before the order is received.

The service level is the probability that all orders will be filled from stock during the replenishment lead time of one order cycle. The service level is the inverse of a stock out percentage. In other words, if the stock out percentage is 5%, then the service level is 95%. This means that during a reorder cycle, there is a 95% probability that a spare part will be available. Figure A-8 shows the impact of factoring safety stock into

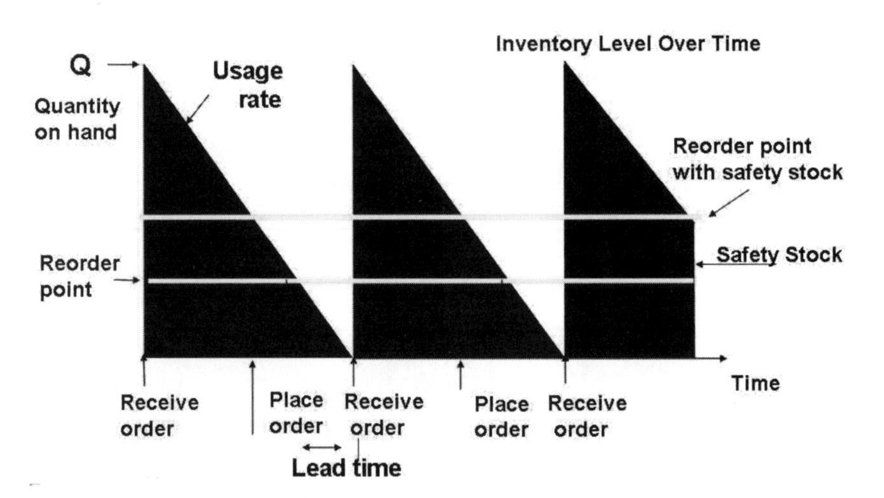

Figure A-8 The MRO Inventory Cycle

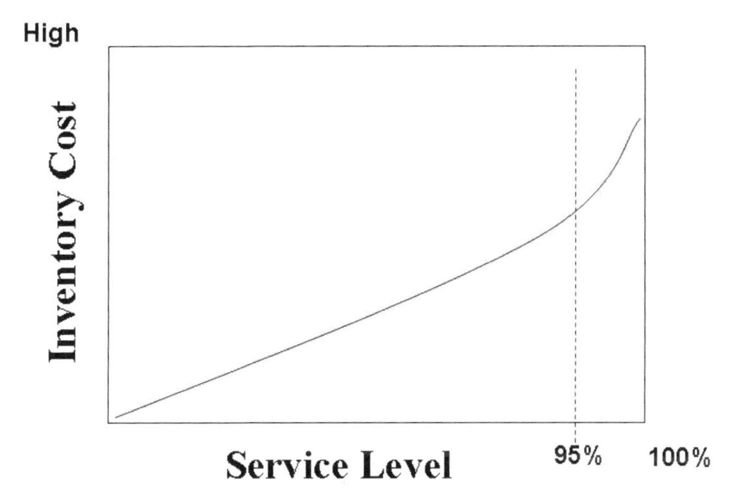

Figure A-9 Cost-Service Trade-off

Figure A-2. The safety stock block is added to the reorder point and a new reorder point with safety stock is established. Adding safety stock requires orders to be placed sooner. Thus, safety stock requires an organization to carry additional inventory. The bottom line is that using safety stock will have an impact on the economic order quantity. However, if the stock out costs is high enough, the safety stock can be justified.

This tradeoff between cost and service is highlighted in Figure A-9. In this figure, as the service level increases towards 100%, the slope of the line increases. The closer that the MRO storeroom gets to a 100% service level, the greater the cost. In fact, stocking enough spare parts in a storeroom to insure there is never a stock out would be incredibly expensive. The only time service levels above 95–97% can be justified are for extremely critical equipment items with a very high cost of downtime.

This tradeoff is one of the main reasons why the ABC classification system is used. As Chapter 2 discussed, spare parts are classified based on their usage, dollar value, holding costs, and stock-out costs. With the ABC classification system, the main focus for EOQ, reorder points, and safety stock is given to the "A" items. These items that will have the largest cost impact on the plant. "A" items are the high value, low utilization spares. "B" items should also be tracked using EOQ, reorder points, and safety stock because these parts are medium value and medium impact. "C" items will likely never come under such scrutiny. They are low-value

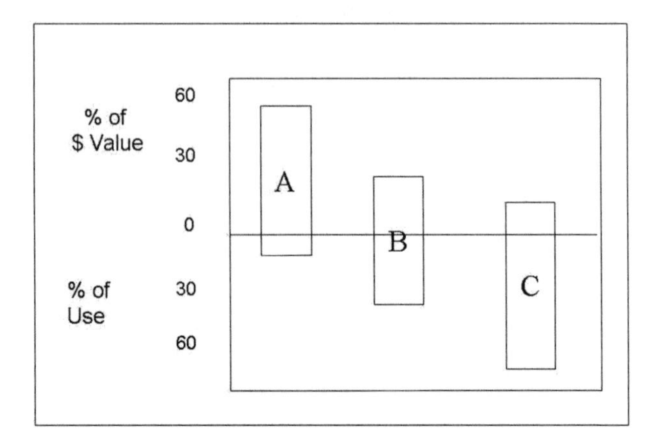

Figure A-10 ABC Classifications

items with very high turnover. They are commonly becoming vendor-stocked items.

The use of economic order quantities, reorder points, and safety stock is important for major spares in MRO storerooms. Using these guidelines, companies should be able to optimize their investment in MRO spares.

INDEX